The Mediaeval Academy of America
Publication No. 84

MEDIEVAL FRENCH BRIDGES

Medieval French Bridges
A History

Marjorie Nice Boyer
York College
City University of New York

THE MEDIAEVAL ACADEMY OF AMERICA
Cambridge, Massachusetts 1976

The publication of this book was made possible by grants of funds to the Mediaeval Academy from the Carnegie Corporation of New York.

Library of Congress Catalog Card Number: 75–36478
ISBN 910956–58–8
Printed in the United States of America

To Carl

Contents

List of Illustrations

(Following p. 80)

Preface

MY INTEREST in medieval French bridges grew out of a preoccupation with travel in that period. Most of my scholarly research has been devoted to a study of medieval travel. I have written on the speed of travel, on carriages and wagons, on the cost of journeying through France, and on other aspects of travel. Because travelers depended on bridges, they were intensely interested in whether or not a particular bridge was passable. Itineraries were planned with regard to bridges, and a new bridge might divert traffic from one route to another. J. J. Jusserand in his *English Wayfaring Life in the Middle Ages* considered it appropriate to include a section on bridges, and I have felt the need to write a book on the subject. There has been no work on the topic which includes the whole of the Middle Ages and all of France, based on an examination of contemporary documentary evidence as well as of the extant bridges.

My thanks go to my friends of the Columbia University Seminar in the Renaissance for their criticism of several chapters read before them, in particular to Professor Paul O. Kristeller for his perspicacious suggestions. The Hatchet Club of the University of Kansas was kind enough to suggest improvements in the first two chapters. The Columbia University Library and the New York Public Library have been especially helpful in providing materials for this work, and the libraries of the University of Chicago and the University of Kansas have also contributed. To the Library of Congress, the libraries of Harvard University and of Duke University, to the Bibliothèque Nationale, and to a number of French departmental archives I am indebted for the privilege of working there, or for loans of rare books, or for microfilm or photostats.

I wish to thank my husband, not only for his generous encouragement of my studies, but also for his patience in reading with the painstaking eye of a practiced editor the whole of the manuscript. Mme A. Lombard of the Ecole

Pratique des Hautes Etudes has been extremely helpful in procuring illustrations. Finally, to Mr. Paul Meyvaert of the Mediaeval Academy I am indebted for his interest and for arrangements leading to the publication of the manuscript, and to Mr. Luke Wenger I am under heavy obligation for his careful attention in preparing the work for the printer.

MARJORIE N. BOYER

Brooklyn, N.Y.
January 24, 1976

1

The Background

IN ALL AGES bridges have had the faculty of exciting the admiration of the beholder. This is certainly true in our own day, which might be considered the great age of bridges, but it was equally true in the Middle Ages. While no medieval bridge could rival in size and technique the structures over the Golden Gate or the Straits of Mackinac, whose long spans and lofty towers place them among the most impressive of man-made achievements, the great medieval stone bridges across the Loire and the Rhone were nonetheless a cause for wonder in their own day. While there are now many more bridges than at any time in the past, no single one of them can match the central importance of the medieval bridge to its community, or fulfill the variety of functions that was taken for granted in the medieval bridge.

A modern bridge is planned as a means of communication or, perhaps, with an eye to military strategy, but bridges in the Middle Ages were something more. It was possible, at various times and places in medieval France, to be born on a bridge in a lying-in hospital, to reside on a bridge and go to market there, to carry on a business, whether retailing or manufacture or both, to attend services in a chapel, and even, in the twelfth century, to hear lectures given by masters of the University of Paris. Often the fortified bridgehead was the entrance to the town. At the time of the joyous entry of a lord whom the town felt obliged to honor, the bridge was decorated. In 1389 when Queen Isabeau de Bavière made a ceremonial entry into her city of Paris, the Grand Pont was brave with green and white hangings. At dusk, as she stood in front of Notre-Dame, she watched an acrobat holding a candle in each hand while he performed on a rope stretched between the "highest tower" of the cathedral and the tallest house on the Pont-Saint-Michel.[1] In England in the reign of King Richard II a joust was held on London Bridge.[2]

1. Jean Froissart, *Chroniques*, in *Oeuvres*, 14, ed. Kervyn de Lettenhove (Brussels, 1872), pp. 11–12.
2. G.C. Home, *Old London Bridge* (London, 1931), p. 92.

If we experience a feeling of awe in viewing the monumental bridges of our day, so also did medieval men when regarding the spans of their time. "Pons ille mirificus" was the description of the Pont-Saint-Bénézet at Avignon given by Robert of Auxerre within twenty-five years of its completion.[3] Similarly, in the medieval period London Bridge was considered a national wonder. This attitude is symptomatic of the fascination evoked by medieval bridges, not only in their own day, but in ours. The Pont-Saint-Esprit, first thrown across the unmanageable Rhone nearly seven centuries ago, rightly commands our respect; and tourists today go into remote parts of France to view the fortified Pont Valentré at Cahors (begun 1308) or the tower of the bridge over the Gave de Pau at Orthez (attested 1254).

The Middle Ages had its own view of bridges, distinct from ancient and from modern views. Both in the medieval period and in antiquity the history of bridges was intertwined with religion, but with striking differences—medieval people assumed that the bridge was under divine protection, while the ancients supposed that the spirit of the stream was likely to be unfavorable to those crossing it, whether by bridge or by ford. For example, Homer in the *Iliad* represents the god of the local watercourse, the Scamander, as attempting to prevent Achilles from destroying the Trojans.[4] With many a pagan this was more than a literary device, and there was an ancient belief that it was wise, on crossing a stream, to throw in a coin to propitiate the spirit of the river. Consequently, the site of London bridge in Roman times has been verified through the ancient coins recovered from the river bed.[5] Money has also been found at ancient fords,[6] and this custom may be behind the modern idea of throwing coins into a fountain while making a wish.

Nevertheless, the Roman connection between bridges and religion seems to have been minor. The most recent opinion holds that it is an error to see a relationship between the Latin word for bridge (*pons, pontis*) and that for priest (*pontifex, pontificis*). *Pontifex* did not originally mean builder of bridges, and the derivation from *pontis* + *facere* is incorrect. Rather, the first element of *pontifex* is derived from a lost Umbrian word, *puntes,* probably meaning certain religious rites involving sacrifices. According to this explanation, the root meaning of *pontifex* was "maker of sacrifices."[7]

3. *Historiens de France,* 12:298.
4. *Iliad* 21, lines 234–294, trans. A.T. Murray (Cambridge, Mass., 1967), 2:425–429.
5. Home, *Old London Bridge*, pp. 5–6.
6. A.Grenier, *Manuel d'archéologie Gallo-Romaine,* Part 2: *L'Archéologie du sol. Les Routes* (Paris, 1934), p. 185.
7. *The Oxford English Dictionary* (Oxford, 1933), s.v. pontifex.

At Rome the Pons Sublicius did have a role in religious rites; on the fifteenth of May the *pontifices* and magistrates gathered on this bridge to see the Vestal Virgins throw into the Tiber twenty-four or twenty-seven puppets of reeds or straw called Argei. The significance of this ceremony was not known to the Romans of the last two centuries B.C. W.W. Fowler believes that it was merely a magical rite to ensure adequate rain for crops, and that the puppets did not represent men to be sacrificed.[8]

Unlike the Romans, medieval people believed that bridges were under divine protection; and unlike the moderns, they did not consider them in a romantic but rather in an allegorical context. A famous omen in medieval literature is the burning of the bridge over the Rhine at Mainz, which according to Einhard presaged the death of Charlemagne. His description of the destruction of this bridge suggests that the event gained in impressiveness from the wonder of the structure itself:

> The wooden bridge over the Rhine at Mayence, which he [Charlemagne] had caused to be constructed with admirable skill, at the cost of ten years of hard work, so that it seemed as if it might last forever, was so completely consumed in three hours by an accidental fire that not a single splinter of it was left, except what was under water.[9]

Bridges lend themselves readily to allegory, a characteristic feature of medieval thought. For example, Gregory the Great tells the story of an allegorical dream vision of a bridge in his *Dialogues.* A soldier lying at the point of death in the great plague year of 590 had a vision of a bridge. It led from this world to the pleasant, flower-scented meadows of heaven, crossing the black and dingy stream of hell, which emitted a cloud of vapor with an intolerable stench. As he crossed the bridge, the soldier felt his foot slip on the narrow passage. The devils below laid hold of him to pull him down, but good spirits seized him and raised him back up on the bridge. Here an historian might have chosen to make some comment on the justice of this picture of a medieval bridge. (Evidence shows that many were indeed foot-spans of minimum width unprovided with protective railings.) Gregory's interests, however, were not so mundane. In answer to questioning he explained that the narrowness of the bridge was due to the words of our Lord, "Narrow is the way that leadeth unto life" (Matthew 7.14).[10]

8. W.W. Fowler, *The Religious Experience of the Roman People from the Earliest Times to the Age of Augustus* (London, 1922), p. 54.
9. Einhard, *The Life of Charlemagne,* trans. S.E. Turner (Ann Arbor, 1960), p. 60.
10. Gregory the Great, *Dialogi Libri IV,* ed. U. Moricca, Fonti per la Storia d'Italia 57 (Rome, 1924), p. 287.

Allegory continued to be popular throughout the Middle Ages. A twelfth-century example of a bridge in this connection appears in Chrétien de Troyes's *Chevalier de la Charrette.* Lancelot, in search of Guinevere, comes to a most unusual bridge:

> If anyone asks of me the truth, there never was such a bad bridge, nor one whose flooring was so bad. The bridge across the cold stream consisted of a polished, gleaming sword; the sword was stout and stiff and was as long as two lances. At each end there was a tree trunk in which the sword was firmly fixed. No one need fear to fall because of its breaking or bending, for its excellence was such that it could support a great weight.[11]

The sword-bridge did indeed support Lancelot firmly during his crossing, but he emerged from the ordeal much lacerated, evidently a symbol of his sufferings for love of Guinevere.

The bridge as an image in allegorical literature need not detain us here; it is a mere sidelight to the history of medieval bridges. Perhaps no less marginal, but of more consequence for the historian because of their influence on popular and scholarly perceptions, are the concerns of the Romantic movement—the love of legends and picturesque ruins, the interest in the magical and the supernatural. That the influence of the Romantic movement is not dead is amply demonstrated by the literature for travelers. It is not the bridges that are still in use, such as the Pont-Saint-Esprit, that find their way into the illustrated guidebooks and provide adornment for travel posters. Over and over again one sees pictures of the Pont-Saint-Bénézet at Avignon, a bridge which not only has been unusable for three centuries—it also retains its abandoned chapel, and its construction was inspired by a saint. The combination is obviously irresistible.

The effect of the romantic approach to bridges, which emphasizes the picturesque at the expense of the functional and useful, is to depreciate the actual role of bridges in the medieval period. Forgotten are such practical matters as the concentrated effort and technical skill required for the construction, the perpetual struggle to preserve the bridge against the efforts of the stream to wash it away, and the simple fact of the benefits provided to travelers.

The romantic view of bridges also shows special affection for a phenomenon which, perhaps surprisingly, is more modern than medieval—the associa-

11. Chrétien de Troyes, *Arthurian Romances*, trans. W.W. Comfort (New York, 1928), pp. 308–309.

tion of bridges with magical and supernatural elements. The *Dictionnaire topographique du département des Vosges* lists for the modern period four bridges named Pont-des-Fées: one in the commune of Bain-les-Bains, one in the commune of Gérardmer, one in the commune of Saint-Amé, and one (appropriately enough in ruins) in the commune of Voivres.[12] Evidence for legends of the devil as a builder of bridges likewise dates from the modern period. The legends tell the story of builders so overcome by the magnitude of their task that they succumb to despair and sell their souls to the devil in exchange for aid in completing the bridge. Invariably the work progresses marvelously with the collaboration of the devil, but the device by which the builder in the end escapes his pact with the devil varies from tale to tale. At Pont Ecumant in the Alps, since the agreement provided that the devil should be privileged to fly away with the first living thing to cross the span, the builder arranged that this should be a dog.[13] At the Pont Valentré (Cahors, France) the legend states that the devil contracted to carry out successfully all the orders of the mason. Accordingly, when the bridge is all but built, the latter assigns to him an impossible task, that of carrying water in a sieve.[14]

Whether or not myths of the devil involved in bridgebuilding were current in the Middle Ages is open to question.[15] Certain it is that in the medieval period saints were associated with bridges, on which chapels were placed. On the façade of a church at Borgo San Donnino in Italy there is a twelfth-century sculpture of the collapse of a wooden bridge. The faithful, hurrying to see the newly-discovered tomb of San Donnino, are shown falling into the river, and the saint is depicted as preserving them.[16] In modern times a number of bridges have been renamed after the devil. For example, a bridge across the Hérault built in the eleventh century by the convents of Aniane and Saint-Guilhem-le-Désert appears in the 1964 edition of the *Guides Bleus* for

12. *Dic. Top. Vosges,* p. 329. Throughout the book all topographical dictionaries of the French departments will be cited in this fasion, with complete information to be found in the bibliography, section III.
13. W.J. Watson, S.R. Watson, *Bridges in History and Legend* (Cleveland, 1937), pp. 35–37.
14. D.Steinman, S.R. Watson, *Bridges and Their Builders,* 2nd ed. (New York, 1957), p. 65.
15. Certainly the theme of the swindled builder was present in medieval Germanic legends, but in the *Edda* it was a giant who had promised to complete Asgard's walls during the winter and was surprised by the coming of summer: A.Wünsche, *Der Sagenkreis der geprellten Teufel* (Vienna, 1905), p. 22.
16. A.K. Porter, *Lombard Architecture,* 4 (New Haven, 1917), plate 30, n.5. C.Ricci, *L'Architettura Romanica in Italia* (Stuttgart, 1925), p. 53.

France as the Pont-du-Diable.[17] Also, in the department of Isère a bridge known in the fourteenth century as the Pons Sancti Hugonis is listed in the modern period as the Pont-du-Diable.[18]

These secondary associations—bridges in allegorical literature, the aesthetics of picturesque ruins, the legends of supernatural bridgebuilders—are not an important concern of the present study, as they were not a very important aspect of medieval attitudes toward bridges. The medieval view of bridges was essentially practical, and it is therefore appropriate to concentrate on how medieval French bridges were built, who built and maintained them, how the necessary funds were acquired, and what the various functions were that the bridges fulfilled.

Medieval bridges exerted an immense influence on routes, commerce, administration, and military affairs. There are innumerable examples in medieval history in which possession of a fortified bridge enabled one faction to dominate the countryside and deny the region to another. The presence or absence of a bridge might make the difference as to whether or not a river formed the boundary of an administrative district. In Dauphiny the impetuous rivers carried away one bridge after another.[19] In the fifteenth century the inhabitants of a town on the Drac requested their transfer for administrative purposes from one district to another, so that they would not have to ferry across that formidable river: "Parce qu'il y a une rivière nommée le Draulx laquelle est dangereuse et perilliouse."[20]

To emphasize the importance of bridges to trade routes one need only remind the reader that the abandonment of the fairs of Champagne by the Italians was facilitated by the opening of the St. Gotthard pass, far to the east of the Mont Cenis, which had hitherto been a favorite of traders journeying from Italy to the Low Countries. The St. Gotthard route encouraged travelers from Italy to pass through the eastern Alps and the Rhineland to reach Flanders, rather than crossing France, and it was made practicable by a new bridge. About the middle of the thirteenth century an enterprising mason put through a roadway, partly supported by beams, in the gorges of Schoellenen

17. G. and F. Monmarché, *France*, Les Guides bleus (Paris, 1964), p. 431.
18. *Dic. top. Isère*, s.v. In the *Dic. top. Aude*, p. 319, there is listed a Pont-du-Diable, a bridge over the canal du Midi, commune of Alzonne, which in 1781 had been called the Pont-des-Romains.
19. The Romanche, a tributary of the Drac, washed out the bridge at Vizille in 1336, 1350, 1388, 1399, 1453, and 1457: T. Sclafert, *Le Haute-Dauphiné au Moyen-Age* (Paris, 1926), pp. 171–172, 442. Vizille (Isère), com. St. Pierre-de-Mesage.
20. Sclafert, *Le Haute-Dauphiné*, p. 166.

and Reuss in the Alps. Where the Reuss made a series of falls through a deep gorge between two cliffs, he built a narrow stone bridge without parapets. It was called the Pont Ecumant, because it was drenched with spray.[21]

Writing on medieval French bridges has been extensive, and yet one agrees with Hubert Gautier, who in the early eighteenth century penned the following statement: "Le sujet des Ponts est assez vaste pour donner de l'occupation aux plus habiles. Jusqu'ici personne n'a traité de cette matière autant qu'il le mérite."[22] As an architect and inspector of the royal bridges and highways for Louis XV, Gautier wrote to disseminate technical information. He stated that bridges were the best thing men had imagined for facilitating commerce, but he complained that prior to his time very little had been written on their construction.

In general, scholarly works on the medieval bridges of France have either relied on only a portion of the evidence or have treated only a section of the field. Most of the books which include treatment of medieval French bridges have been written from the architectural or engineering point of view, with a few composed from the popular and picturesque angle. Most of the works on bridges compiled for laymen confine themselves to one chapter on medieval bridges together with a few pictures.[23] In the nineteenth and twentieth centuries many of the extant medieval bridges of France have been measured and described by Dartein and Degrand, and sketched and photographed by William Emerson and Georges Gromort and by Paul Séjourné.[24] This has been an important contribution to our knowledge of the subject, but the visual evidence still available in our own day fails to give an accurate and rounded picture of what French bridges were like in the Middle Ages. The addition of contemporary documentary evidence is essential. There have also been histories of individual bridges, as of the Pont-Saint-Esprit and of the

21. C. Gilliard, "Problèmes d'histoire routière. I. L'Ouverture du Gothard," *Annales, Economies, Sociétés, Civilisations* 1 (1929), 177–182.

22. H. Gautier, *Traité des ponts ou il est parlé de ceux des Romains et de ceux des Modernes*, first published 1716, cited from 4th ed. (Paris, 1765), 1, preface. See also J.R. Perronet, *Description des projets et de la construction des ponts de Neuilly, de Mantes, d'Orléans & autres*, 2 vols. (Paris, 1782–1783).

23. Steinman and Watson, *Bridges and Their Builders*. C.S. Whitney, *Bridges: A Study in Their Art, Science and Evolution* (New York, 1929). E.B. Mock, *The Architecture of Bridges* (New York, 1949).

24. F. de Dartein, *Etudes sur les ponts en pierre remarquables par leur décoration*, 1 (Paris, 1912). E. Degrand, J. Résal, *Ponts en maçonnerie*, 2 (Paris, 1888). W. Emerson, G. Gromort, *Old Bridges of France* (New York, 1925). P. Séjourné, *Grandes Voûtes*, 6 vols. (Bourges, 1913–1916).

Pont-de-la-Guillotière at Lyons,[25] but each is on one span only, and these accounts appeared in local journals many years ago. The consequence of this fragmentary approach to medieval bridgebuilding is that the subject has been distorted by a disproportionate emphasis on those bridges that have happened to survive and on those facets of the subject, like the bridgebuilding brotherhoods, which have caught the fancy of writers. The purpose of the present study is to recount the history of the bridges of medieval France, as it can be established from the available evidence, with particular attention to bridgebuilders—those responsible for construction and maintenance. The bridges treated will be those built in the medieval period in the region now constituting modern France.

The evidence on which to base our conception of medieval bridges consists partly of the extant structures and partly of documents—government orders, legal records, and especially expense accounts. There are in France today a number of bridges dating back to medieval times, as at Albi, Lyons, Céret, Espalion, and Montauban. However, the extant structures are far from giving a correct idea of what the numbers and condition of bridges must have been like in the Middle Ages. With few exceptions they are tucked away in remote corners of France. In France today all the existing medieval bridges are in stone. However, in the eleventh and twelfth centuries the vast majority of bridges were of wood, and one of the reasons for contemporary admiration of the Pont-Saint-Bénézet at Avignon was that the immense structure was built of stone.

Metamorphosis from wood to stone or (after a flood) from stone to wood is only one of the vicissitudes to which bridges are subject in the course of their history. By their function they are exposed not only to the usual weathering of all construction, but also to the erosion of the stream they cross. Again and again we read that such and such a bridge was washed out by a flood. Then, too, bridges, like other buildings, are subject to changes in taste and in the requirements of the age. Most of the medieval bridges were widened in the nineteenth century. The most common of alterations is probably the removal of the buildings upon them in accord with early modern taste—especially the seventeenth- and eighteenth-century preference for the un-

25. L. Bruguier-Roure, ed., *Cartulaire de l'oeuvre des église, maison, pont et hôpitaux du Saint-Esprit, Mèm, Acad. Nîmes,* ser. 7, Annexes to vols. 12 (1889), 13 (1890), 14 (1891), 16 (1893), 17 (1894). M.C. Guigue, "Notre-Dame de Lyon; recherches sur l'origine du pont de la Guillotière et du grand Hôtel-Dieu et sur l'emplacement de l'hôpital fondé à Lyon, au VIe siècle, par le roi Childebert et la reine Ultragothe," *Mém. Lyon* (1874–1875), pp. 227–365.

cluttered. The attitude of this later period was entirely opposed to that of Guillaume Paradin, who wrote his *Mémoires de l'histoire de Lyon* in the sixteenth century. He remarks with approval that bridges are not merely stone structures: "They are erected with so much artifice that, being on these bridges, you are unable to judge whether you are on a bridge or streets continuing from the town, so much is the whole ornamented and clothed with magnificent houses."[26] Almost all French bridges surviving from the Middle Ages have long since lost their houses and towers. Fourteenth-century documents provide that there shall be three towers garrisoned by royal troops on the bridge at Montauban, but the fortifications are now gone, and we know that one of the towers was demolished in the sixteenth century because it blocked the view from the bishop's palace.

Alterations in the visual form of the bridge make it all the more important to consult the documentary evidence to determine the history of medieval bridges. Chroniclers took the catastrophic view of bridges, so that they usually provide the reader with a number of references to the destruction of spans by fire, flood, or neglect. Indeed, we are sometimes left wondering when a span could have been rebuilt in time to be carried away by still another inundation. Documents are more likely to be helpful. Chartularies of religious establishments meticuously record their rights with regard to bridges, and much can be learned about the Pont-de-la-Guillotière from the records of the abbey of Ainay at Lyons. Lawsuits have been responsible for some of our largest collections of documents, as for example at Avignon and at Pont-Saint-Esprit. A large number of surviving governmental documents tell of towns granted permission to levy taxes for bridgebuilding and maintenance; but the most copious and useful information on the details of construction comes from financial accounts, such as those of Albi and of "the bridge of Orleans and the hospital [of Saint Anthony] situated on this bridge."[27] A useful collection of information is the series of *Dictionnaires topographiques* of the departments, which provide the earliest known references to many of the bridges.

The history of bridges, like other history, is to an extent dependent on chance as to what we have to tell. Accidents, floods, and nineteenth-century ideas of modernization have powerfully affected the survival of medieval bridges into our own era. For some periods, such as the Merovingian and

26. G. Paradin, *Mémoires de l'histoire de Lyon* (Lyons, 1573), p. 120.
27. "Du pont d'Orleans et de l'ospital estant sus y cellui pont," Archives du Loiret, CC 920.

Carolingian, the paucity of information on bridges reflects a general scarcity of written sources. Nevertheless, a judicious use of the extant documentation, together with a study of surviving bridges and archeological evidence, makes it possible to present a balanced picture of the history of medieval bridges in France.

In the study that follows three major periods are distinguished in the history of medieval French bridges. In the first period, which lasted until the collapse of the Carolingians, bridges were considered a governmental responsibility exercised primarily for military purposes. Carolingian legislation continued the Roman practice of requiring local inhabitants to build and maintain specific bridges. Few stone bridges were built in this period.

In the second period, lasting from the end of the tenth century until at least the middle of the thirteenth century, when so much of the regalia passed into private hands, bridges too became the domain of individuals and local institutions. In other words, many bridges became private property, and in much of France the government had little or no responsibility for bridges. Nevertheless, construction in stone was undertaken on a large scale, and at least in some areas of France bridges came to be seen as a pious work, like hospitals and churches worthy to receive the donations of the faithful.

In the last period, from about the middle of the thirteenth century to the end of the Middle Ages, the government once again took a more active role. Bridges came to be looked upon chiefly as public works, serving the inhabitants of a region and facilitating commerce and travel. Various tolls and levies were approved by the government to provide more adequate funds for maintenance. The period was marked by significant improvements in construction techniques.

I

FROM THE ROMANS THROUGH THE CAROLINGIANS

2

The Bridge as Military Adjunct

THE ROMANS seem to have viewed bridges as subordinate to a road system planned to facilitate troop movements. In Rome under the principate the layout of highways was determined for the whole empire. Where convenient, this planning included the construction of bridges, as on major routes like the Via Domitia stretching across southern Gaul to Spain.

The Roman achievements in bridgebuilding could be magnificent, but in Gaul the most impressive monument of this sort was the Pont-du-Gard, which was originally built as part of an aqueduct and only later converted into a bridge (see Fig. 6). The Romans seem never to have constructed a bridge across the mighty Rhone[1] except at Arles, and that was a bridge of boats, apparently with a masonry arch next to the shore. In lieu of bridges the Romans at certain points made use of permanent fords, in some cases paving them or sinking a wooden flooring to facilitate passage. In antiquity there was

1. It has been maintained that there was a Gallic bridge across the Rhone near the present town of Pont-Saint-Esprit: C.L. Jullian, *Histoire de la Gaule* (Brussels, 1964), 2:231, n. 7. The evidence is a passage in Orosius concerning the defeat of Bituitus. king of the Averni, by Consul Quintus Fabius Maximus in 121 B.C.: "Realizing that the one bridge over the Rhone River was too small for him to lead his troops across, Bituitus had another constructed by chaining together small boats over which he spread boards and fastened them down" (*Seven Books against the Pagans: The Apology of Paulus Orosius* 5.14.2, trans. I.W. Raymond [New York, 1936], p. 229). Orosius lacks repute as an historian. He is also by far the latest of the ancient authors to write on this battle, his *Books against the Pagans* dating from the early fifth century. Livy, Pliny the Elder, Strabo, and Valerius Maximus, who wrote within 200 years of the battle, do indeed note the triumph of Quintus Fabius Maximus over Bituitus, but they do not mention a bridge: Livy, *Periochae* 61, trans. B.O. Forster, 14 vols. (London, 1919–1959); Pliny the Elder, *Naturalis Historiae* 7.166, trans. H. Rackham, W.H.S. Jones, D.E. Eichholz, 10 vols. (Cambridge, Mass., 1938–1962); Strabo, *The Geography* 4.2.3, trans. H.L. Jones, 8 vols. (Cambridge, Mass., 1948); Valerius Maximus, *Factorum dictorumque memorabilium libri novem* 9.6.3, ed. J. Kapii (London, 1823).

a paved ford across the Claim at Poitiers, remains of which were found in 1897–1898 in rebuilding the Pont de Rochereuil. The substructure consisted of large rocks, which were surmounted by paving stones.[2]

It has been said that "whereas the Romans were road-conscious but were quite prepared to cross rivers by ford, the men of the Middle Ages were essentially bridge-conscious."[3] The great esteem in which bridges were held is already evident among the ancient Gauls. The earliest datable information on Gallic bridges is supplied by Roman sources. Caesar began his campaigns in Gaul in 58 B.C., and his *Gallic Wars* mentions a bridge on the Aisne, and three on the Seine—one at Melun and two at Paris, one on each side of the Ile-de-la-Cité.[4] There were several bridges over the Allier, all of which Vercingetorix broke down in 52 B.C. The Romans were marching along one bank of the river, the Gauls on the other, the latter preventing the former from repairing any of the spans. Caesar tricked Vercingetorix into marching on in surveillance of the major part of the Roman troops. Then the rear guard quickly put a floor back on the uninjured piles and successfully crossed the Allier.[5]

The sites of several bridges mentioned by Caesar are in dispute. Thus most scholars equate Genabum with Orleans, but Napoleon III assumed it was the same as modern Gien.[6] Also, there is a dispute as to whether the bridge across the Loire used by Dumnacus in fleeing from Caesar was at Saumur or at Ponts-de-Cé.[7] The historian of the latter place, A.A. Bretaudeau, insisted that there was a bridge there in the first century B.C. and that it represented the most direct and convenient route from Angers to Poitiers.[8] However, Camille Jullian in his *Histoire de la Gaule* preferred Saumur, on the ground that Saumur is nearer Poitiers than is Ponts-de-Cé.[9]

2. A. Richard, *Histoire des comtes de Poitou, 778–1204* (Paris, 1903), 1:324. See also A. Grenier, *Manuel d'archéologie gallo-romaine* (Paris, 1934), 2:185–187 (hereafter cited as Grenier, *Manuel*).
3. *The Cambridge Economic History*, ed. M. Postan and E.E. Rich (Cambridge, Eng., 1936), 2:147. C.T. Flower, ed., *Public Works in Mediaeval Law*, 2 (London, 1923), p. xix.
4. Caesar, *De bello gallico* 2.5.6; 7.58.5, 6, ed. St. George W.J. Stock (Oxford, 1898).
5. *De bello gallico* 7.34.3.
6. *De bello gallico* 7.11.6. See Stock's edition, p. 325. See also A. Collin, "Existe-t-il des vestiges apparents d'un pont dans le lit de la Loire en face de Gien-le-Vaux?" *Mém. Orléannais* 19 (1886), 253–290; soundings disclosed no remains of a Roman bridge.
7. T.R.E. Holmes, *Caesar's Conquest of Gaul*, 2nd ed. (London, 1911), p. 831.
8. A.A. Bretaudeau, *Histoire des Ponts-de-Cé*, 2 vols. in 1 (Angers, 1901–1903), pp. 7–8.
9. Jullian, *Histoire de la Gaule*, 2:554, n. 4.

Roman itineraries—vases listing post stations, the Antonine and Jerusalem itineraries, and the Peutinger Table—in enumerating the principal towns on routes mention a few named after bridges: Pompertuzat (Haute-Garonne) near Toulouse, the *Mutatio Ponte Herarum* or relay station on the post road between Bellegarde and Arles,[10] the *Pons Dubis* [Pontoux (Saône-et-Loire)] on the way from Chalon-sur-Saône to Augst via Besançon, and the *Pons Scaldis* [Escaupont (Nord)].[11] The *Pons Saravi* mentioned in the Peutinger Table has been identified with Fénestrange (Meurthe).[12] The bridge at Arles must date from a period later than the first century, when two lists, inscribed on silver vases, of the post stations on the road between Cadiz and Rome assume that the Rhone must be crossed by boat near this point.[13] Evidence for the bridge at Arles comes from Ausonius writing in the fourth century and from Cassiodorus in the sixth, as well as from a mosaic discovered at Ostia,[14] representing this bridge of boats.

Archeology indicates that there were Roman bridges at Autun, at Sully-sur-Loire, at Saint-Thibaut (Cher), at Limoges, at Périgueux, and at Chalon-sur-Saône (where the structure replaced a Gallic ferry).[15] Especially the Via Domitia and its branches were provided with bridges. In the department of the Hérault, from Vidourle on the east to Lake Capestrang on the west, there were at least six bridges.[16] In Roman times Auxerre, Saintes, Sens, Narbonne, Besançon, and Carcassonne had bridges, the last being a timber span.[17] The

10. H. Gröhler, *Über Ursprung und Bedeutung der französischen Ortsnamen*, 2 vols. (Heidelberg, 1913–1933), 1:148 (hereafter cited as Gröhler, *Ortsnamen*). E. Desjardins, *Géographie historique et administrative de la Gaule Romaine*, 4 vols. (Paris, 1876–1893), 4:33, 34.
11. Desjardins, *Géographie*, 4:143, 127. Gröhler, *Ortsnamen*, 2:147. Pontes near Amiens has been accepted by scholars as being identical with the modern Ponches (Somme), but Gröhler denies that the Roman name can be the ancestor of the present designation on the grounds that such an evaluation violates the rules for the development of the Picard dialect (*Ortsnamen*, 1:277).
12. Desjardins, *Géographie*, 4:133.
13. Desjardins, *Géographie*, 3:478, n. 2.
14. Grenier, *Manuel*, 2:495, nn. 1, 4.
15. For Autun see H. de Fontenay, *Autun et ses monuments* (Autun, 1889), pp. 204–205. For Sully-sur-Loire and Saint-Thibaut see R. Dion, *Le Val de Loire: Etude de géographie régionale* (Tours, 1934), pp. 293–295. For Limoges see F. de Verneilh, "Architecture civile du moyen-âge. Construction des ponts," *Annales arch.* 20 (1860), 100. For Périgueux see *Dic. top. Dordogne*, s.v. For Chalon-sur-Saône see A. Bartier, "La Traversée de la Saône à Chalon," *Annales de Bourgogne* 26 (1954), 129.
16. Grenier, *Manuel*, 2:190.
17. For Auxerre and Sens see M. Quantin, "Histoire de la rivière d'Yonne," *Bull. Yonne*

two bridges at Paris were of wood in the fourth century, according to an oration, *Misopogon* (enemy of the beard), by Julian the Apostate (361–363).[18] The most famous of Roman timber bridges was Caesar's across the Rhine (see Fig. 1), but the extant Roman bridges are of stone. Some of these are yet to be seen in southern France. Still in use are the bridges at Sommières (Gard) (see Fig. 2), the Pont Julien (Vaucluse), the Pont Flavien near Saint-Chamas (Bouches-du-Rhône), and the Boisseron (Hérault) on the Bénovie.[19]

The Romans, in conformity with their lack of appreciation of bridges, named very few towns after bridges, preferring to perpetuate the fame of illustrious men. On the other hand, very many Gallic towns took their names from bridges, thus demonstrating an interest in bridges which carried over into the Middle Ages. The Celtic equivalent of "pons" was "briva," and there is a long list of place names in Gaul with "briva." Brives-la-Gaillarde (Corrèze) was mentioned by Gregory of Tours as "Briva Curretia vicus," the village with the bridge on the Corrèze.[20] The name Briva Isarae, the bridge over the Isara, was translated and became Pontoise, the bridge over the Oise. Samarobriva (Amiens) meant the bridge over the Somme. Other places conserving the memory of bridges are noted by Albert Grenier in his *Manuel d'archéologie gallo-romaine.*[21] He assumes that most bridges indicated by place names with

39 (1885), 431, 440. For Saintes see E. E. Viollet-le-Duc, *Dictionnaire raisonné de l'architecture française du XIe au XVIe siècle*, 7 (Paris, 1864), 231. For Narbonne see Sidonius Apollinaris, *Poems and Letters*, trans. W.B. Anderson (Cambridge, Mass., 1936), 1:286, Carmen 23, line 44. See also A. Blanc, ed., *Le Livre de comptes de Jacme Olivier, marchand narbonnaise du XIVe siècle* (Paris, 1899), 2,1:lii. P. Marnotte thought the old bridge at Besançon was Roman: "Mémoire sur la voie romaine qui traversait Besançon," *Acad. Besançon* (24 August 1852), pp. 121 ff., 136. Grenier (*Manuel*, 1:173, n. 4, 174; 2:263) mentions other Roman bridges: Pontchartrain in Jouars (Seine-et-Oise), Pontchartrain (Orne), Pontchartrain in Brézolles (Eure-et-Loir), Pont de Rêmes in Florent (Marne), Pont-du-Roi (Seine-et-Oise, com. Tigeray), Pontvy (Savoie), and elsewhere.

18. *The Works of Emperor Julian*, trans. W.C. Wright (London, 1913), 2:421.

19. J. Sautel and L. Imbert, *Les Villes romaines de la vallée du Rhône* (Avignon, 1926), pp. 82, 102. Grenier, *Manual*, 1:568. Grenier doubts that the old bridge at Vaison is Roman (2:676).

20. Gregory of Tours, *Historia Francorum* 3.7.10, ed. W. Arndt, MGH SSrerMerov 1 (Hanover, 1885).

21. Grenier, *Manuel*, 2:264. For a list of such names consult the Appendix. As the final component of place names, "briva" appears in the Peutinger Table in "Gabris," on the road from Bourges to Tours at the crossing of the Cher. "Gabris" was a derivation of Caro-briva (bridge on the Cher), today Chabris (Indre). Salbris (Loire-et-Cher) means Salera, i.e., Sauldre plus "Briva" or Sauldre-bridge. See also Gröhler, *Ortsnamen*, 1:142.

the element "briva" must have been built before the end of Gallo-Roman times, because very few were constructed in the Merovingian era.[22]

Bridgebuilding in the Roman Empire was one of the *munera*, obligations laid on communities. Inhabitants were required to provide labor and town councils to construct bridges. The Pont Flavien near Saint-Chamas (Bouches-du-Rhône) was built by a local magnate, just as prominent citizens donated temples, theaters, and amphitheaters, but such munificence had already declined in the third century, and the Frankish conquest of the fifth century brought in a climate of opinion unfavorable to the construction of new bridges or the maintenance of old. The Frankish idea of law as customary inhibited the building of bridges at new sites, because nothing novel and different could be required of the inhabitants of a neighborhood. The Merovingians retained tolls on certain bridges, but in some cases their collection was made over to private persons or to abbeys. The Merovingian idea of kingly power was limited. As the responsibilities of the monarch were small, so his expenses were modest. The Merovingians allowed the collection of the Roman land tax to lapse, apparently feeling that it was a kind of tribute paid into the royal treasury. As such, it was bad form to require it of Frankish subjects, and the approved method of replenishing the king's hoard was by raids into the territory of neighbors.

Evidence for only a handful of bridges has survived from this period: at Paris, Pompierre (Vosges), Pontaudemer (Eure), Ponthion (Marne), Pont-de-Ruan (Indre-et-Loire), Pontlieue (Sarthe), Pont-sur-Seine (Aube), Sens, Verdun, Brives-la-Gaillarde (Corrèze), and Charenton.[23] Bridges at Paris are mentioned in two sixth-century sources. Fortunatus's life of St. Leobinus, bishop of Chartres (d. ca. 536), describes a bridge at Paris with a typically medieval characteristic, that is, it was covered with houses. The life of St. Leobinus relates how he was invited to Paris to celebrate mass at Easter. In the middle of the night fire broke out in the houses constructed along the length of the bridge, and although water from the river was thrown on the fire continu-

22. Grenier, *Manuel*, 2:262.

23. For Pontaudemer see *Dic. top. Eure*, s.v. For Pont-de-Ruan and Pont-sur-Seine see A. Longnon, *Géographie de la Gaule au VIe siècle* (Paris, 1878), pp. 286, 333. For Pontlieue (Sarthe) see *Dic. top. Sarthe*, s.v. For Sens see Quantin, "Histoire d'Yonne," p. 440. For Charenton see A. Poncelet, "Relation originale du prêtre Idon sur la translation de S. Liboire à Paderborn," *Analecta Bollandiana* 22 (1903), 146–172. See also A.J.V. Le Roux de Lincy and L. M. Tisserand, *Paris et ses historiens aux XIVe et XVe siècles* (Paris, 1867), p. 226, n. 1. For the others see Gregory of Tours, *Historia Francorum* 3.26; 5.11 (17); 6.12 (19); 6.23 (32); 8.33; 10.9.

ously, the flames did not die down, and they threatened the town. The story continues that at this point the prayers of the bishop were efficacious in putting out the fire.[24] The passage refers to the houses as "hanging," perhaps meaning that they projected from the stone piers.

Gregory of Tours (538–594) twice mentions Parisian bridges, on the occasion of a fire on the bridge over the southern branch of the Seine, and in connection with the capture of a man fleeing across the city bridge, who was unfortunate enough to break his leg when his foot slipped between two planks forming the bridge.[25] The bridge was obviously in a poor state of repair. Although the Latin would admit of the interpretation that the bridge was only two planks wide, such a narrow width is unlikely for a Parisian bridge, one at least of which was already covered with houses. Evidently the references are to the Petit Pont and to the Grand Pont, that is, bridges across the smaller and larger branches of the Seine. The former mention explains that the fire broke out in a house nearest the southern gate of the town, while in the latter case, the man fled "per pontem civitatis." "The city bridge" was the designation used for the Grand Pont on numerous occasions from the tenth to the fourteenth centuries.[26]

24. Fortunatus, *Vita sancti Leobini*, ed. B. Krusch, MGH AA 4,2:79: "Edax ignis exiliens domos pendulas quae per pontem constructae erant exuere coepit et non solum ex vicino fluvio incessantur aqua superfusa non adquievit, sed etiam civitati proximus civibus ut universa consumeret magnum timorem incussit."
25. Gregory of Tours, *Historia Francorum* 6.23 (32): "Cumque per pontem urbis fugiret, elapso inter duos axes, qui pontem faciunt, pede, effracta oppressus est tibia." See also *Historia Francorum* 8.33.

 Gregory mentions a bridge at Tours, but it was not a bridge across the Loire, for it was supported by only two boats. Apparently the bridge served the purpose of facilitating embarcation on the river. In the later Middle Ages the word bridge (*pons* or *pont*) was frequently used for a walkway protruding into the river, and that is the probable meaning of "pontem" in the following passage: "Interea ingressi in fluvium super pontem qui duabus lintribus tenebatur, navis illa, quae Leudastem vehebat, dimergitur, et nisi natandi fuisset aminiculo liberabutus, eum sociis forsitan interisset" (*Historia Francorum* 5.49). This may be translated: "Meanwhile having embarked on the river [passing] over the bridge, which was supported by two boats. . . ." However, O.M. Dalton translates: "Meanwhile they had embarked on the river above the bridge supported by two boats, and the vessel bearing Leudast sank, so that he would probably have perished with his companions, if he had not saved himself by swimming" (*History of the Franks* [Oxford, 1927], 2:223). See also the translation of E. Brehaut, *History of the Franks* (New York, 1916), p. 137.
26. A few of the references to the "pons civitatis," i.e., the Grand Pont: *History of the Franks* 8.33, trans. Dalton, 2:357–358. R.C. Lasteyrie du Saillant, ed., *Cartulaire génèrale de Paris* (Paris, 1887), p. 4, no. 2, in 558, "A ponte civitatis." Ibid., p. 48, no. 34, in 829, "in ponte Parisius." *Gallia Christiana*, 7, Instrumenta, no. 18, col.

Despite a scarcity of bridges in the Merovingian kingdom, at least two place names would imply that there was more than one at a site. In 715 the present town of Pontaudemer was referred to as Duos Pontes.[27] As for Pont-sur-Seine, throughout the Middle Ages the name was always given in the plural. Under the date of 574, Fredegarius called it Duodecim Pontes,[28] and "Locus qui XII Pontes super Sequanama dicitur" was the term used in 804 in a chartulary.[29] The reason for the name is uncertain; no town in Gaul was large enough in Merovingian or Carolingian times to require twelve bridges.

More can be learned about bridges in Carolingian than in Merovingian times, because the literary remains are much more extensive. Copies of abbey privileges mention bridges, the royal capitularies regulate tolls and repairs, and the annals record the vicissitudes of the spans during the invasions of the Northmen. In accordance with the Frankish concept of law as customary, the capitularies continually insist on precedent. Charlemagne and Louis the Pious reiterate the good old customs of their ancestor Pepin. It is the old tolls that are to be collected, and the bridges are to be rebuilt where they formerly stood. There is, however, an altogether exceptional admonition in a capitulary of Louis the Pious, dated 821, referring to twelve bridges on the Seine which the *missi dominici* are to direct local people to rebuild. They are not to listen to vain arguments as to whether the bridges should be rebuilt where they formerly stood, for they must be constructed where they are now necessary.[30] Presumably the site of the new bridge would have been in the same general locality as the old, for it was the people of that region who were responsible for maintaining it.

The Carolingian capitularies provided that certain classes of people were to be exempt from tolls, such as pilgrims going to Rome, persons traveling from one of their houses to another, officials proceeding on royal business,

17, in 909, "pontem iam dictae urbis." S. Luce, "Pièces inédites relatives à Etienne Marcel et à quelques-uns de ses principaux adhérents," *Bibl. Ecole Chartes* 21 (1859–1860), 87, in 1362, "per de subtus archam Pontis Parisius."

27. L. Delisle, L. Passy, eds., *Mémoires et notes d'Auguste Le Prevost pour servir à l'histoire du département de l'Eure*, 2 (Evreux, 1864), 548.

28. *Dic top. Aude*, s.v. Fredegar, et al., *Chronica. Vitae sanctorum*, ed. B. Krusch, MGH SSrerMerov 2 (1888), #71, p. 112. Another instance is found in the accounts of the *hôtel* of Philip VI for 1337, where Pont-Sainte-Maxence is called "Pontes Sancte Maxencie": R. Fawtier, "Un compte de menues dépenses de l'hôtel du roi Philippe VI de Valois pour le premier semestre de l'année 1337," *Comité des Travaux hist.* (1928–1929), p. 207.

29. J.J. Bourassé, *Cartulaire de Cormery*, Mém. Touraine 12 (1861), p. 10.

30. *Capitularia regum Francorum*, ed. A. Boretius, MGH Capit 1:301.

and persons going to the king's palace.[31] Certain attitudes toward tolls on bridges foreshadow later medieval practice. For example, when Louis the Pious (ca. 820) provided that no man who had made a bridge should be forced to pay toll, it suggests that these were probably local inhabitants, who also in the later Middle Ages were insistent on free passage across bridges. The same capitulary provided that if anyone should repair a bridge "ex propriis facultatibus," he could not on that account increase the charges.[32] There was evidently much trouble with over-zealous toll collectors, and the kings keep insisting that no boatman passing under the bridge was liable for toll for the boat. Nevertheless, this practice was recognized at Paris in 759,[33] and in the later Middle Ages it was customary to collect tolls on goods passing under as well as over bridges.

The capitularies do not make a necessary connection between receiving the tolls and maintaining the bridge. It is possible that in an age when there was not much trade and, in addition, local people and royal officials were exempted from paying toll, the income from a particular bridge would have been very inadequate to maintain it. The legal provision for the upkeep of roads and bridges in Carolingian times was founded on a statement in the Theodosian Code of ca. 439–that persons living in the neighborhood were responsible for public works, including bridges, and that no one was exempt from this obligation.[34] These provisions were expressed in capitularies of Pepin, Charlemagne, Louis the Pious, and Louis II. Louis the Pious in 818–819 instructed the *missi dominici* to associate with themselves, for the repair of bridges, the local bishop and count.[35] Louis's capitulary of 823–825 insisted that the old bridges should be rebuilt on the same sites.[36] Both the repeated injunctions against the collection of unjust tolls and the reiterated orders to repair bridges seem to imply a difficulty in the enforcement of royal orders. In 829 Louis the Pious ordered men who had flouted his orders to

31. MGH Capit 1:32, 124–125.
32. MGH Capit 1:294–295.
33. F. Lot, "Mélanges Carolingiens: Suite II. Le Pont de Pîtres," *Le Moyen-Age* 18 (1905), 138. J. Tardif, *Monuments historiques [Carton des rois] 528–1789* (Paris, 1886), p. 638, #57 bis. A Parisian bridge was attested in 836, when monks transporting the body of Saint Liborium, bishop of Le Mans, to Paderborn were met "ultra pontem" by the bishop of Paris, who conducted them honorably into his church: Poncelet, "Relation du prêtre Idon," p. 165.
34. Theodosius, *Libri XVI cum constitutionibus Sirmondianis et leges novellae ad Theodosianum pertinentes* 15.3, 6, ed. T. Mommsen and P.H. Meyer (Berlin, 1905).
35. MGH Capit 1:288.
36. MGH Capit 1:306–307.

repair bridges to come to his palace to explain their negligence.[37] In 850 Louis II ordered that old bridges should be restored, and in 854 Charles the Bald commanded that they should be rebuilt by those holding the *honores* from which the bridges had earlier been built or restored.[38]

In the Carolingian period the Roman idea of bridges as a factor important to military operations continued to be held. Chapter 10 of the Capitulare Aquisgranense (801–813) provides for the supplying of equipment to Charlemagne's army. The chapter reads in part: "And each count shall save two parts of the fodder of his county for the army's use, and he shall have good bridges, good boats."[39]

The ninth century saw the rise of a new and distinctively medieval concept of the role of bridges in warfare. The modern idea is to build bridges to increase one's own troop mobility or to destroy them to prevent their use by the enemy. The medieval concept was somewhat different. It was to fortify and garrison the bridge to prevent the enemy from crossing over it or passing under it. The invasions of the Northmen were largely carried out by boat, and they found the navigable rivers of France inviting avenues by which to penetrate to the heart of the country. They ascended the Loire, and they went up the Seine into the county of Burgundy. To meet this threat Charles the Bald conceived the idea of building bridges, not only to facilitate crossing streams, but also to deny their use to the Northmen. When in 862 he rebuilt the bridge at Trilbardou (Treix) on the Marne, he was successful in preventing the Vikings from passing farther along the river.[40] The same year he ordered his men, gathering for the annual May assembly, to bring with them workmen and carts for the construction of a fortified bridge of wood and stone near his palace of Pîtres, very probably at Pont-de-l'Arche.[41] The Northmen seem to have interrupted work on the bridge, and it remained ineffective as a bar to their progress up the Seine. Charles the Bald also planned to rebuild the ancient bridges at Charenton beyond Paris and at Auvers-sur-Oise (Oise), arrondissement of Pontoise. He did build a "castrum" in connection with the bridge

37. MGH Capit 2:16.
38. MGH Capit 2:84, 87, 277.
39. MGH Capit 1:171, translated in S.C. Easton and H. Wieruszowski, *The Era of Charlemagne: Frankish State and Society* (Princeton, 1961), p. 139.
40. *Annales Bertiniani*, ed. G. Waitz, MGH SSrG (Hanover, 1883), p. 57. Trilbardou is Treix (Haute-Marne).
41. Lot, "Mélanges Carolingiens: Suite II," p. 1 and n. 3. The history of the abbey of Saint Florent near Saumur credits Charles the Bald with having built a fortified bridge on the Loire. Presumably this was at Ponts-de-Cé.

at Auvers, but it was unsuccessful in stopping the Northmen–the garrison surrendered the bridge tower in 885.[42]

That fortified bridges could be effective in stopping the Norse fleets was proved both at Treix on the Marne and at Paris on the Seine. Although it is known that there had been two bridges at Paris in the first century B.C. and in the sixth century A.D., Charles the Bald found it necessary to rebuild the Grand Pont sometime between 861 and 870.

The sorry record of the later Carolingians in defending the country against the Northmen was repeated in the summer of 885 when a Neustrian army, assembled to prevent the enemy from ascending the Seine, failed to defend the fortified bridge at Pîtres. It was another matter, however, when the invaders reached Paris. The defense of the city was organized by Bishop Joscelin and by Odo, son of Robert the Strong, count of the march of Neustria. It was the heroic defense of the city by Odo, contrasted with the supine ineptitude of Charles the Fat, that resulted in the deposition of the Carolingian ruler and the crowning of Odo in 887. Abbo, monk of St. Germain-de-Près, has left us an epic poem written in 897 about the siege of Paris (885–887).[43] Abbo tells us that on 24 November 885 Northmen to the number of 40,000 arrived on more than 700 vessels, so that their boats covered the river for two leagues. They demanded permission to pass and go upstream. The Parisians refused. Two days later, on 26 November, the Northmen attacked the tower at the entrance to the Grand Pont on the right bank of the Seine. Protected by ditches, the defenders were able to hold out despite the missiles. On 31 January 886 one group of Northmen assaulted the tower, and two others took to their boats to attack the bridge. The Seine disappeared beneath the Norse bucklers. These onslaughts, too, were beaten back. Furious, the Northmen filled three boats with brush, set the boughs and twigs alight, and, walking along the bank, towed the vessels downstream to set the bridge afire. However, the boats lodged against masonry (presumably the piers), and the Parisians on the bridge were able to extinguish the flames with water from the river and to capture the boats. After this, the Grand Pont seems not to have been molested.

It was next the turn of the Petit Pont. A catastrophe for the defenders

42. *Annales Xantenses et Annales Vedastini*, ed. B. de Simson, MGH SSrerG (Hanover, 1909), pp. 57–58.

43. Abbo, *Le Siège de Paris par les Normandes. Poème du XIe siècle*, ed. and trans. H. Waquet (Paris, 1942). See also E. Favre, *Eudes, comte de Paris et roi de France (882–898). Annales de l'histoire de France à l'époque carolingienne* (Paris, 1893), pp. 35 ff.

occurred during the night of 26 February, when the Seine rose and carried away the middle section of the bridge to the south, leaving a dozen defenders isolated in the tower on the Left Bank. The Northmen drove a cart full of burning materials against the tower, forcing the garrison out on the ruins of the bridge. Their surrender and massacre followed. Finally Charles the Fat arrived with an army to succor the city, but he merely bribed the Northmen to leave, giving them permission to go up the river to Burgundy. As the Parisians continued to refuse to allow the invaders to pass under the Grand Pont, the Northmen dragged their boats around and sailed up the Seine in defiance of the defenders of the city.

Many contradictory statements have been written about the bridges of Paris in this period, and the controversy is still continuing. The authenticity of some of the early documents is suspect. In particular, doubt centers on the genuineness of a thirteenth-century charter purporting to be a copy of an 861 donation to the bishop of Paris by Charles the Bald of "the greater bridge" (majorem pontem) he had built in the territory of Saint-Germain-l'Auxerrois.[44] The date contains a contradiction between the calendar year and the regnal year, and the term "greater bridge" (rather than Magnus Pons) seems unique in the documents. Also, no bridge is known to have been built in the territory of Saint-Germain-l'Auxerrois. Charles's construction of the Grand Pont was formerly thought to be corroborated by a statement of the chronicler Adon which mentioned Charles's building of a "wonderfully strong bridge across the Seine," but Ado was probably referring to the bridge at Pîtres.[45] Nevertheless, it seems certain that Charles the Bald either built or more probably repaired the Grand Pont, both because of a document of 909 in which Charles the Simple confirmed the gift by his ancestor Charles of the city bridge to the bishop of Paris and because in the thirteenth century jurisdiction on the Grand Pont belonged to the cathedral chapter.[46] Perhaps the 861 document on the Grand Pont as we have it was a thirteenth-century forgery based on an authentic document deliberately suppressed. Or, perhaps, the original document having been lost, some member of the chapter undertook to supply the lack, accounting for the clumsiness of the thirteenth-century copy with its incorrect date and unconvincing details.

There have been disagreements as to the site of the bridge of Charles the

44. Lasteyrie du Saillant, *Cartulaire génèrale*, pp. 62 ff.
45. Lot, "Mélanges Carolingiens: Suite II," p. 138. "Ex Adonis episcopi Viennensis chronico," *Historiens de France*, 7:55.
46. B.E.C. Guérard, ed., *Cartulaire de l'église Notre-Dame de Paris* (Paris, 1850), 4,1:461. *Gallia Christiana*, 7, Instrumenta, col. 16 ff.

Bald. Some writers have been convinced that his bridge constituted a third Parisian bridge, and that "majorem pontem facere" meant to construct a bridge bigger than the Grand Pont, not the Petit Pont.[47] The arguments of Hurtaut and Magny and of Jaillot for the third bridge are based on a series of misconceptions. For example, it was built, according to Jaillot, because the Grand Pont, covered with houses, was not defensible. However, in the Middle Ages such bridges were a part of town defenses and the scene of military actions, as for example at Orleans during the English siege in 1428. In addition, an unfortified bridge would have been an avenue leading the Northmen right to the walls of Paris and would have been a military liability. Such a bridge would surely have figured in Abbo's epic.

The two chief choices advanced for the site of Charles the Bald's bridge are those of the present-day Pont-au-Change and Pont-Notre-Dame. Those writers who have hypothesized that the ninth-century bridge was the predecessor of the Pont-Notre-Dame have assumed that it was the same as the "Planches de Mibray," to which in the thirteenth century there was "une ruelle" leading from either side of the Seine.[48] However, an investigation of this point shows that the Planches de Mibray merely led to mills in the Seine and not from one bank to the other, as will be discussed below.[49] Another reason for assuming that the bridge of Charles the Bald was at the site of the

47. P.T.N. Hurtaut and Magny, *Dictionnaire historique de la ville de Paris* (Paris, 1779), 4:94. Le sieur Jaillot, *Recherches critiques, historiques et topographiques sur la ville de Paris, depuis les commencemens connus jusqu'à présent* (Paris, 1782), 1:163 ff. A. Berty, "Recherches sur l'origine et la situation du Grand Pont de Paris, du Pont-aux-Changeurs, du Pont-aux-Meuniers et de celui de Charles le Chauve," *Rev. arch.* 12 (1855), 203. Borrelli de Serres, "L'Agrandissement du Palais de la Cité sous Philippe le Bel," *Mém Paris* 38 (1911), 9 ff. Borrelli de Serres discusses various writers' views on the location of the oldest bridges at Paris and of the bridge of Charles the Bald. See also Charles Duplomb, *Histoire générale des ponts de Paris* (Paris, 1911–1913), 1:87–91. Jean Guérout has suggested that the Grand Pont was only built on its present site in the reign of Robert the Pious (996–1031) and that the bridge was constructed between 1111 and 1142: "Le Palais de la Cité à Paris, des origines à 1417. Essai topographique et archéologique," *Mém. Féd. Paris* 2 (1950), 130–140. Anne Lombard-Jourdan considers the bridge built by Charles the Bald to have been originally a defensive construction, a barrage, which was later used as a bridge. It was located, she thinks, on the site of the present-day Pont-Notre-Dame: *Paris: Genèse de la "Ville." La rive droite de la Seine des origines à 1223* (Paris, 1976), pp. 37–38.

48. P.H.J.F. Géraud, *Paris sous Philippe le Bel d'après un manuscrit contenant le Rôle de la Taille imposée sur le habitants de Paris en 1292* (Paris, 1837), p. 301. Borrelli de Serres, "L'Agrandissement," p. 9.

49. See below, p. 140.

Pont-Notre-Dame was that the latter is located directly north of the Petit Pont. To intellectual heirs of an eighteenth-century preference elegance the attractiveness and efficiency of straight streets have appeared so obvious that these authors have assumed that in the Middle Ages there must necessarily have been a direct passage across the Ile-de-la-Cité from the Right Bank to the Left. However, medieval people seem to have been less interested in town planning. Many towns of the period developed in a haphazard manner with crooked streets, and Paris was no exception.

It is known that at least from the thirteenth century the Grand Pont was not only north but also west of the Petit Pont and very near the site of the Pont-au-Change. Inside the right-bank abutment of the Pont-au-Change a pier and two arches with very definite Roman characteristics were discovered in 1855. These findings were hailed as remains of Charles the Bald's bridge, but the dating has been contested.[50] The persistence of tradition would also indicate that the bridge of Charles the Bald must have been at the same location. In the thirteenth century the jurisdiction of the cathedral chapter extended along the northern branch of the Seine from the Ile-Sainte-Marie (now the Ile-Saint-Louis) down to and including the Grand Pont but not extending beyond.[51] This must have been a legacy from Charles the Bald's gift of his bridge to the bishop of Paris in the ninth century. In the Middle Ages to build a new bridge was a much more difficult undertaking than to repair an old one, from the juridical point of view. It has already been pointed out that the Carolingian capitularies commonly insist on the reconstruction of a bridge on the old site, and when in 821 Louis the Pious ordered that bridges were to be rebuilt where they were needed, not necessarily in their former location, his tone was defensive.[52] Evidently the use of a new location was very rare in Carolingian times. To construct a bridge on a new site required the consent of anyone owning that part of the river or its banks. The history of medieval French bridges is full of instances of compensation given to those having rights at the new location. The probability, therefore, is that as long as there were only two bridges at Paris, from ancient times to 1378, the solitary

50. Théodore Vacquer, "Lettre à M. l'éditeur de le *Revue archéologique* sur la découverte d'une partie du Grand Pont de Paris bâti par Charles le Chauve," *Rev. arch.* 12 (1855), 502–507. See below, p. 80.

51. L. Tanon, *Histoire des justices des anciennes églises et communautés monastiques de Paris suivie des registres inédits de Saint-Maur-des-Fossés, Sainte-Geneviève, Saint-Germain-des Près et du registre de Saint-Martin-des-Champs* (Paris, 1883), p. 153.

52. See above, n. 32.

bridge across the northern branch of the Seine was at almost the same site, about where the Pont-au-Change now is.

There has been some misunderstanding, too, as to how the Northmen could have found the bridges of Paris so formidable an obstacle, especially after the Petit Pont had been swept away.[53] The collapse of later bridges choked the stream with debris and stopped all navigation. Even when a medieval bridge was functional, it was common for it to have a single arch reserved for navigation; the other arches were too narrow, the flow of water was inadequate, or they were blocked with mills, fishponds, or rocks. At Paris in the thirteenth and fourteenth centuries the navigable arch of the Grand Pont was called "the arch of the bridge of Paris" or the master arch of Grand Pont.[54] The presence of armed men on the bridge could be a powerful deterrent to river traffic. The successful blocking of Norse passage up the Seine by the bridges of Paris found a parallel when in 1016 Cnut, invading England, found London bridge, then a wooden structure, so formidable an obstacle that he preferred to find a way for his ships through the marshes.[55] Similarly, in 1263 a mob congregated on London Bridge and, armed with whatever missiles came to hand, was able to prevent the unpopular Queen Eleanor of Provence from passing under the bridge. The queen gave up the attempt and returned to safety in the Tower.[56]

The cynical lassitude of Charles the Fat at the siege of Paris resulted in his dethronement and the substitution of the Robertians as rulers in 887. The eclipse of the Carolingians was only temporary at this time, and it was not until 987 that they were permanently replaced by the Capetians. The collapse of the Carolingian empire meant that there was no longer a central government powerful enough to insist on the building and maintenance of bridges. They passed into private hands, and as the counts appropriated other powers and possessions of the monarchy, they also asserted their claim to collect tolls on bridges. The disorder attendant on the Norse invasions occasioned the neglect of repairs, and in the tenth century few bridges appear in the records. For the period between 751, the coronation of Pepin, and 887, the deposition of Charles the Fat, the capitularies, annals, and other documents must be supplemented by the evidence collected in the topographical dictionaries of the

53. Jaillot, 1:167–168.

54. A. Terroine, L. Fossier, eds., *Chartes et documents de l'abbaye de Saint-Magloire* (Paris, 1966), 2:491, $324. Luce, "Pièces inédites," *Bibl. Ecole Chartres* 21 (1859–1860), 87.

55. G.C. Home, *Old London Bridge* (London, 1931), p. 10.

56. Home, *Old London Bridge*, p. 55.

French departments to determine the names of bridges. For the period between 888 and 987, evidence from capitularies, annals, and the like is largely absent, and the topographical dictionaries, compiled for the most part from chartularies, become the chief repository of information. Since the bridges appear in the chartularies as incidental references in boundary descriptions, the historian is obliged to infer the location of bridges from passages such as the following: "Pont-de-Barret (Drôme,) commune of the canton of Dieulefit. Locus qui prius dictus est Savenna et modo qui dicitur ad Pontem. 965."[57] Presumably the want of sources for this period reflects an actual scarcity of bridges, and only half as many bridges are attested for the period 888–987 as for the years 751–887. Such a result is perhaps to be expected at a time when bridges became private property.

57. *Dic. top. Drôme*, p. 281. For other bridges of the period, consult the Appendix.

II

FROM CA. 1000 TO CA. 1250

BRIDGES IN PRIVATE OR CORPORATE POSSESSION

3

The Bridge as a Pious Work

The Flowering of an Idea

IN THE PERIOD from the eleventh to the thirteenth century a number of new mechanisms were developed to assure the building and maintenance of bridges. The bridge came to be seen as a pious work, the suitable object of charity, and along with this new concept came a new institution, the *opus pontis* (the "fabric of the bridge"), which constituted an endowed corporation. These ideas continued to be honored throughout the Middle Ages despite the fact that many people at all times viewed bridges in quite another light and although beginning in the thirteenth century charity made proportionally much less of a contribution to bridge construction than it had earlier.

The concept of the bridge as a pious work appeared in response to a particular set of conditions. Under the Carolingians the ultimate responsibility for bridge construction had rested with the emperor, who ordered the counts to assure maintenance by means of corvées performed by residents of an area, who were then allowed free passage over the bridge. With the debacle of Carolingian power the obligation of residents to maintain the bridge in their vicinity lapsed. In the tenth century the ownership of bridges, as well as of watercourses, roads, and fishing rights, and the authority to collect certain taxes, were appropriated by lay and ecclesiastical lords. The response of the people of what is now France to the collapse of central authority was different in the North and in the South. Between the Loire and the Rhine fuedalism developed as a means of providing "peace and justice." In northern France, between the year 1000 and the early twelfth century, about the only people with sufficient power and financial resources for bridge construction were lay and ecclesiastical lords, and almost all bridges built at this time were constructed by them. However, in the South the attempt to provide order took the form of the Peace of God movement, with ideas clearly related to that of the bridge as a pious work.

The notion of the bridge as an object of charity is documented in the eleventh century. The concept was clearly borrowed from the church. The clergy attempted to move into the breach caused by the breakdown of central government and to supply, or to encourage others to supply, some of the former services. Some bishops became temporal rulers, and in southern France clergymen organized what later was called the Peace of God movement.[1] This activity began in Aquitaine in the tenth century and was especially important in southern France before spreading to Catalonia, Lombardy, and Germany. The objectives were peace; the elimination of violence against churchmen and the poor; an end to the robbery of clergy and peasants and to destruction of vineyards, mills, and farm animals; the restoration of church lands; and the prohibition of castle building. The chief means to enforce compliance was the calling of assemblies of the local clergy and notables to bring public opinion to bear on the recalcitrant and the threat of excommunication against those violating their vow to maintain the peace.

There are important parallels between the inception of the Peace of God movement and the history of the bridge as a pious work. In each case the reward was to be a heavenly one. Like the Peace of God movement, the inspiration for making the bridge an object of charity was evidently originally clerical. The concept was similarly anti-militaristic, for it forbade the fortification of bridges, and it provided for the poor, demanding that passage across bridges should be gratuitous for all, not merely for local residents. Any money donated should be for the love of God, and on occasion the decision to build a bridge was made in an assembly of notables.

Records exist of three bridges planned in assemblies in southern France, two in the first half of the eleventh century and one, at Toulouse, in the early twelfth century. The convents of Aniane and Saint Guilhem-le-Désert (then called Gellone), in the presence of laymen of the region, agreed sometime between 1031 and 1048 to build a bridge over the Hérault at the Gouffre-Noir. Each abbot and his monks promised to pay a master mason for building half the bridge. No motive for the undertaking was given in the document, but two of the provisions usual for a bridge built as a pious work were included: no charge was to be collected for passage across the bridge, and fortifications were prohibited. It has been suggested that the motive for its construction may have been to divert traffice from the ancient Via Domitia to a route passing the convents.[2]

1. H. Hoffmann, *Gottesfriede und Treuga Dei* (Stuttgart, 1964), p. 4. The earliest references are to *pactum pacis, constitutio pacis, restauratio pacis,* etc.
2. C. Devic, J. Vaissète, *Histoire générale de Languedoc*, ed. E. Mabille, E. Barry,

At Albi, too, it was an assembly that planned the construction (ca. 1035) of a bridge in the time of Bishop Amelius, who had been instrumental in introducing the Truce of God into the Albigeois.[3] The purpose in building the bridge was "the common improvement of the town and the utility of the whole Albigeois."[4] Besides the assembly there are only two connections with the concept of the bridge as a pious work: the provision that passage should be free to all and the mention of the fabric of the bridge (*opus pontis*). Our information is derived from a document drawn up by the canons of Saint-Salvi to safeguard their rights in the bridge. They explained that they had been urged by the bishops of Albi, Nîmes, Rodez, and Cahors, by the viscount of Albi, and by the citizens of the *cité* and the *bourg* of Albi to donate to the fabric of the bridge the land on which the bridge was to be constructed and which had been a ferry port. For the canons of Saint-Salvi the gift of the port involved the loss of the ferry tolls. Consequently, if tolls should ever be charged on the bridge, the canons wished to register their prior claim to any such monies. This document was singularly ineffective in achieving its purpose, and when at a later date tolls were collected, it was not the canons of Saint-Salvi who received them, but the viscount of Albi.

At Toulouse, again, the preliminary moves toward construction of the Pont-de-la-Daurade, like the bridges at the Gouffre-Noir and at Albi, were made in an assembly. The documents recording this event must have been written between the accession to power of Alphonse Jourdain, count of Toulouse, in 1119 and the death of Prior Raymond of the Daurade in 1130.[5] Here the idea of the bridge as a pious work was made explicit: passage across the bridge was to be free, and no one was to give anything unless spontaneously and for the love of God. Count Alphonse Jourdain granted to "God and the fabric [of the church] of the Blessed Mary, to Raymond the prior and all the monks of that place, present and future, as well as to the abbots of Cluny and Moissac and Saint Stephen, protomartyr, and to Saint Saturnin and to all

---

E. Roschach (Toulouse, 1872–1905), vol. 5, col. 393, #CLXV. F.J.V. Mortet, ed., *Recueil de textes relatifs à l'histoire de l'architecture et à la condition des architectes en France, au moyen âge, XIe–XIIe siècles* (Paris, 1911), p. 109. See also A.R. Lewis, *The Development of French and Catalan Society, 718–1050* (Austin, 1965), p. 396.

3. Hoffmann, *Gottesfriede*, p. 90.
4. *Gallia Christiana*, 1, Instrumenta, no. 6, p. 4: "pontem fieri super Tarinum, in alodio beati Salvii, ad communem villae meliorationem, & totius Albereis utilitatem."
5. J.H. Mundy, R.W. Emery, B.N. Nelson, eds., *Essays in Medieval Life and Thought Presented in Honor of Austin Patterson Evans* (New York, 1955), p. 184, n. 5.

the men of Toulouse, as well burghers as citizens, that they should hold in perpetuity the bridge they want between the hospital of the Blessed Mary and the Vivarias."[6] The count promised to proceed against anyone who presumed to interfere with this gift. In return the monks of Saint Mary of the Daurade were to sing a mass annually for the repose of the souls of the count and his family. Whether this assembly was effective in accomplishing its purpose is unknown; it is certain only that the Pont-de-la-Daurade was standing by 1181.[7]

The idea of the bridge as a pious work was also current in the eleventh century in the Loire valley. One of the earliest and much the fullest statement in the eleventh century of the bridge as a pious work comes from this region at about the same time that the bridge at Albi was being planned. Between 1031 and 1037 Eudes II, count of Blois, announced construction of a bridge across the Loire at Tours for the benefit of his soul. In an age when his peers were lavishing gifts on churches and undertaking spectacular pilgrimages to Rome or Jerusalem, Count Eudes during most of his life was not conspicuously pious. Only in his later years was he generous to the convent of Marmoutier. The abbey evidently had a hand in inducing Eudes to build the bridge. His charter was composed in the Marmoutier style and preserved into the eighteenth century in the convent's archives, and indeed the monastery stood to benefit from the bridge by increased accessibility and therefore pilgrimages. The charter states that according to the parable of the talents, Count Eudes of Blois, having been called to a great station in life, must needs render to God a greater accounting. Obligated to do great things, but not prepared to do more at present, he had determined on "something memorable, fit for the profit of posterity and on that account pleasing to God."[8] With this in mind he ordered a bridge built across the Loire, near the town of Tours, for there he had known many to perish in the swift flow of the river. Upon consummation of this great work, in order to assure him of his heavenly reward, he decreed that passage across the bridge should be free for all men, whether local residents or pilgrims or merchants, whether pedestrians or

6. G.Catel, *Histoire de comtes de Tolose* (Toulouse, 1623), p. 196.
7. P.Wolff, *Histoire de Toulouse* (Toulouse, 1958), p. 63.
8. E.Martène and U.Durand, *Thesaurus novus anecdotorum* (Paris, 1717), 1:175–176. A. Luchaire, *Les Premiers Capétiens (987–1137)* (Paris, 1901), p. 57. S.Lecoanet, "Charte de Eudes II, comte de Blois, 1033–1037," *Bull. Touraine* 35 (1969), 244–247. C. Chantelou, *Cartulaire tourangeau et sceaux des abbés* (Tours, 1879), p. 17. Interestingly enough, Dom Chantelou attributes the count's order to build the bridge to his affection for his town of Tours–*oppidi amor.*

horsemen, whether poor men or rich men, whether traveling with carts or with loaded or unloaded horses.

Some forty years later (ca. 1080) the same idea, that of saving lives in flood time, appeared as the motive behind the building of two stone bridges over the Gers by Raymond Guiraud, canon and provost of Saint-Saturnin of Toulouse. He pitied those drowned when the raging waters of the flooded Gers sank ferry boats. Raymond Guiraud went further than had Eudes in equating bridge construction with other forms of charity; he diverted to bridgebuilding sums collected for the relief of the poor.[9] Similarly, in 1150 the stone bridge over the Saône at Lyons was built with sums left over from money collected to relieve victims of the plague.[10]

The concept of the bridge as a pious work was most fully developed in southern France, especially in the valley of the Rhone, but also in the Loire valley. In medieval France only in these two regions were there bridges constituting an *opus pontis* and only in the Rhone valley were there bridgebuilding brotherhoods. However, the general idea of the bridge as a pious work was widely diffused. In the eleventh century in parts of Europe as far removed as Spain and Scandinavia, to build a bridge was a good deed aiding a soul in purgatory, and it merited listing as a charity along with the redemption of captives and donations to monasteries.[11] In England before 1122 London Bridge owned lands,[12] and constituted an *opus pontis.* In northern France the frequent floods of the Hem moved Abbot Peter of Andres to rebuild the bridge at Ausque on the road near the castle of Saint-Omer and the town of Thérouanne (Pas-de-Calais).[13] His reasons were to protect poor, weak pilgrims and pedestrians and to eliminate the delays and dangers suffered by all. Therefore at some time before 1178 the abbot ordered the master mason, other masons, and workmen who had built the church to reconstruct the bridge at Ausque of hard stone hewn at a distant quarry. Passage over the bridge was free, explained William of Andres, for the abbot's motive in building it was love of God and that his poor should have rest.[14] (Here should be

9. Mortet, *Recueil*, pp. 260–262.
10. L. Bruguier-Roure, *Les Constructeurs des ponts au moyen-âge. Récits légendaire ou historiques* (Paris, 1876), p. 36.
11. A.U. Arleta, ed., *Cartulario de San Jaun de la Peña* (Valencia, 1963), 2:201. J. Brønsted, *The Vikings*, trans. Kalle Skov (Baltimore, 1965), pp. 201–203.
12. G.C. Home, *Old London Bridge* (London, 1931), p. 18.
13. Thérouanne (Pas-de-Calais), arr. St. Omer.
14. Mortet, *Recueil*, p. 392.

noted the interest of the abbey of Andres, like that of Marmoutier, in encouraging pilgrims to visit the convent.)

The pious motive, in addition to reasons of defense, utility, and profit, was instrumental in accelerating the tempo of bridgebuilding. The tenth century had marked a low point in bridge construction in France. The years between 987 (the accession of Hugh Capet) and 1100 provide us with the names of five times as many bridges as the period between 888 and 987; and the topographical dictionaries, instead of furnishing almost all the names of new bridges, give us only half of them. Chronicles and especially documents attesting property rights furnish details on the rest.

The twelfth century inaugurated the age of the great stone bridges which persisted throughout much of the thirteenth century. The same energy which produced the great cathedrals of Paris, Chartres, Laon, Bourges, Beauvais, and Amiens also produced some of the mightiest achievements in bridge construction of the entire Middle Ages, including the most famous of them all, London Bridge and the Pont-Saint-Bénézet at Avignon. There was a conviction that the widest and most difficult rivers of France could and should be spanned, and that these bridges should, where feasible, be constructed in the more durable medium of stone rather than wood. In the course of the twelfth century the Loire was spanned at Saumur before 1162, Orleans before 1176, and Beaugency before 1160–1182.[15] Over the Rhone bridges were being built at Avignon and Lyons in the late twelfth century, and funds were being collected in 1249 to rebuild the old stone bridge at Vienne.[16] The demand for bridges resulted in new sites and the multiplication of spans within the same town. For example, at Tours we now first hear of the Pont-de-Cher (1172–1178) as well as the Pont-de-Loire.[17]

About the middle of the twelfth century there are increasing signs of a change in the type of person financing bridges. They continued to be built by

15. For Saumur, see E. Berger, ed., *Recueil des actes de Henri II, roi d'Angleterre et duc de Normandie, concernant les provinces françaises et les affaires de France*, 2 vols. in 4 (Paris, 1909–1927), 2:365. See also V. Mortet and P. Deschamps, eds., *Recueil de textes relatifs à l'histoire de l'architecture et à la condition des architectes en France, XIIe–XIIIe siècles* (Paris, 1929), p. 106. For Orleans see A. Collin, *Le Pont des Tourelles à Orléans, 1120–1760* (Orleans, 1895), pp. 390–393. For Beaugency see G. Vignat, ed., *Cartulaire de l'abbaye de Notre-Dame de Baugency*, 1 (Orleans, 1879), 87, 124.

16. M.-C. Guigue, ed., *Cartulaire lyonnais*, 2 vols. (Paris, 1885–1893), 1:576, #454. C.U.J. Chevalier, *Choix de documents historiques inédits sur le Dauphiné* (Montbéliard, 1874), p. 3.

17. Berger, ed., *Recueil Henri II*, 2:84. In 1172, 1173, and 1178.

individuals rather than by the government, but there was by then sufficient diffusion of wealth so that it was practical to collect donations from many individuals rather than to persuade a great lord, lay or ecclesiastical, to finance a bridge. The idea of the bridge as a pious work was enthusiastically taken up by laymen. The bourgeoisie, who in eleventh- and early twelfth-century documents at Albi and Toulouse had been mentioned after bishops and abbots as among those collaborating in planning bridges, now begin to appear either as prime movers or, as at Saumur, as cooperating with knights for this purpose. An act of Henry II, king of England (1154–1189) and count of Anjou, dated 1162, states that in his time the burghers and knights of Saumur "pro remedio animarum suarum" built a wooden bridge across the Loire. The king, rejoiced at so good a work, thanked the knights and burghers, but the abbot of Saint-Florent complained that the bridge had put an end to the abbey's ferry, and this entailed a serious loss of money for the convent. Under these circumstances Henry II presented to the abbey the income from tolls on the condition that the convent should undertake to rebuild one arch in stone every year until the bridge should be completely reconstructed. The document details tolls to be collected on the bridge. Charges were to be paid on merchandise, but not on objects not for sale; on animals to be sold, but not on property belonging to monks, clerks, and servants of lords, or on objects unconnected with commerce. Because the knights and burghers had made so good and useful a work, they were exempted from payment on crossing the bridge with their own property; but they were urged when making their wills to leave sums to the bridge.[18]

Donations financed the greatest bridges of this period—those at Avignon and the Pont-de-la-Guillotière at Lyons—as well as many smaller spans, as, for example, at Saintes and La Rochelle. At these two places bridges were built in a short time "by the alms of the faithful with great labor and expense," according to letters of King John of England, sealed in 1202.[19]

In our own times by far the most famous medieval bridge in France is the Pont-Saint-Bénézet at Avignon (see Figs. 7–9). There are only four arches remaining of the original twenty-two, and there is a deserted chapel perched on the cutwater of one of the piers. The promoter of the bridge became a saint, and there is even a song about it. The Pont-Saint-Bénézet is a press agent's dream. Superior publicity has been enjoyed by this bridge during much of its

18. Berger, ed., *Recueil Henry II*, 1:365. See also P. Marchegay, ed., *Archives d'Anjou* (Angers, 1843–1864), 2:155–156.

19. Mortet and Deschamps, *Recueil*, pp. 194–195.

history and especially at its inception. All we can say of most medieval French bridges is that they must have been built at some time prior to such and such a date. For example, we know that the great bridge at Blois must have been constructed before 1078, because at that date Etienne, count of Blois, presented the Benedictines of Pontlevoy with two mills belonging to him located at the bridge.[20] At Avignon we have the life of Saint Bénézet, written for collectors for the bridge with a careful eye to the prejudices of prospective donors, as well as documents going back to 1180.[21] Both the legend of Saint Bénézet and the chronicle of Brother Robert of Auxerre (d. 1212) give the date of Bénézet's arrival at Avignon as 13 September 1177.[22] Further information is supplied by beatification proceedings, for Bénézet's prestige was such that they were undertaken in the lifetime of those who had known him. However, even at the time of the flowering of the idea of bridgebuilding as a pious work, initiating the Pont-Saint-Bénézet across the mighty Rhone was insufficient to merit sainthood. According to the beatification proceedings, it seems to have been his miracles of faith healing that prompted Bénézet's designation as a saint.[23]

The legend of St. Bénézet affirms that as a twelve-year-old boy, keeping his mother's sheep, he had a vision which required him to proceed to Avignon and to build a bridge across the Rhone. Arriving at the town on the day of an eclipse of the sun,[24] he harangued the people to persuade them to carry out the divine mission. After meeting with jeers from the populace, threats of physical violence from the bishop of Avignon, and skepticism from the povost, he performed the miracle of raising and throwing into the Rhone, to found the first pier, a stone which would have required thirty men to move. This triumph assured an enthusiastic response to his requests for donations. Witnesses at the investigation leading to Bénézet's beatification testified that they had personally known him, had seen him place the first stone, and had accompanied him on his journeys to collect funds. Daily he performed miracles, curing many of blindness, lameness and other ills.

The idea of building great stone bridges was in the air at this time in

20. L.C. Bergevin, A. Dupré, *Histoire de Blois*, 2 vols. (Blois, 1856–1867), 1:406.
21. *Acta Sanctorum*, April, vol. 2, 2nd ed. (Paris, 1865), pp. 256 ff.
22. *Historiens de France*, 12:298.
23. *Acta Sanctorum*, April, 2:258.
24. However, the eclipse of the sun took place on 1 September, not in 1177 but in 1178. See F. Lefort, "Le Premier Pont construit sur le Rhone à Avignon," *Trav. Acad. Reims* 71 (1881–1882), 387–392. See also *L'Art de vérifier les dates*, 3rd ed. (Paris, 1783–1787), 1:73.

France and in England, but considering the difficulty of the site with the powerful current of the Rhone, one can understand the incredulity of the provost, who asked why Bénézet thought he could build a bridge where neither "God nor Peter nor Paul nor even Charlemagne nor another" had done it.[25] Bénézet earnestly began the task of collecting funds for the bridge, journeying as far as Burgundy to do so. He organized the *opus pontis Rhodani*, already the usual type of corporation to administer bridge property. In the case of a cathedral it was the chapter that administered the *opus* or fabric, in the case of a convent the abbot or prior and monks, but here Bénézet instituted a novelty. He started a society of brothers of the bridge, laymen described as *fratres donati*, that is to say, lay brothers who had given themselves to the fabric of the bridge. There is no evidence that they ever took the three vows of chastity, poverty, and obedience. The brothers' chief function was to act as a corporation to solicit, purchase, and hold property for the bridge. The parallel between the language used in donations to a convent and to an *opus pontis* is clear, if one compares the wording of the act of the count of Toulouse with that of the earliest document concerning the bridge at Avignon. At Toulouse the count granted to "God and the fabric of [the church of] the Blessed Mary, to Raymond the prior and all the monks of that place . . . the bridge they want. . . ."[26] At Avignon in April 1181 a garden was sold "to God and to the fabric of the bridge across the Rhone, and to Brother Benedict, initiator and minister of its fabric and to the other brothers of the fabric."[27]

The charitable solicitations of the brothers of the bridge were remarkably successful. They not only collected funds sufficient for the erection of the bridge in only eleven years, but in 1187 they were able to begin the process of buying out the proprietors of the port. Over a period of twenty-nine years the *fratres pontis* were able to accumulate the necessary money and to persuade the "lords of the port" to part with their rights to tolls.[28] There are extant many documents confirming purchase by the brothers of houses,

25. *Acta Sanctorum*, April, 2:257.
26. Catel, *Histoire*, p. 196: "Do et concedo Deo & Beatae Mariae fabricatore & Raimundo Priori, & omnibus senioribus eiusdem loci . . . ut habeant in perpetuum Pontem quem voluerint. . . ."
27. P. Pansier, "Histoire de l'ordre des frères du pont d'Avignon," *Annales Avignon* 7 (1920–21), 34: ". . . Vendimus et venditionis titulo tradimus deo, et operi pontis Rodani, et fratre Benedicto ejusdem operis inceptori et ministro, ceterisque ejusdem operis fratribus. . . ."
28. Pansier, "Histoire," pp. 35, 37, 42–45, 47–49, 53, 55.

gardens, and vineyards many of them at Rognonas (Bouches-du-Rhône), arrondissement Arles, where the parish church was presented to them in 1213 by the bishop.[29] In the fourteenth century brothers worked in the vineyards belonging to the fabric, and they may have done so in the thirteenth.

The crucial function of the brothers, nevertheless, was the collection of funds. The solitary bit of evidence that Bénézet or the *fratres pontis* ever had anything to do with the construction of the bridge is the statement that he laid the first stone. Indeed, all the documents show Bénézet and the brethren of the bridge as exclusively concerned with its financial affairs. As to the actual building of the bridge, the local authorities, the consuls and the bishop, attended to this. In January 1185/6 it was they who settled the question of the tolls to be charged on the bridge.[30]

The society of the *fratres pontis* consisted of both brothers and sisters, as shown by one of the documents mentioning Sister Aldiarde as procurator of the fabric.[31] Apparently the members were already living communally in August 1187 when the "domus operis pontis Rodani" is mentioned.[32] By 1196 documents note their refectory and by 1208 a dormitory with a porch connecting the two rooms.[33] In 1187 the bishop allowed the brethren to have a chapel, a chaplain, and a cemetery. The prior of the association was to be chosen by the members at an election to which the bishop had been invited. Also, for disciplinary reasons the latter might remove the prior, and the bishop's advice was to be taken by the brothers in choosing a chaplain.[34]

The three-month siege of Avignon by the forces of King Louis VIII in 1226 brought ruin to the bridge.[35] This event, however, was not nearly so destructive to the society of the brothers of the bridge as a change in organization made in 1241–1261. Bishop Zoën, instead of allowing the brethren to elect one of their own number as prior, appointed to the position an outsider, Raymond Hugonis, canon of St.-Paul-de-Mausolée. In 1304 one of the brothers, Pierre Raymbert, petitioned the bishop to order the prior ("if there is one") to live with the brothers, to wear their habit, and to provide for them. Formerly the brethren had distributed alms two or three times a week to the poor coming to the house to beg, giving the remainder to the fabric of

29. Pansier, "Histoire," pp. 40–41, 52, passim.
30. *Acta Sanctorum*, April, 2:260.
31. Pansier, "Histoire," p. 53.
32. Pansier, "Histoire," pp. 37–38.
33. Pansier, "Histoire," pp. 42, 50.
34. Pansier, "Histoire," pp. 37–38.
35. Matthew Paris, *Chronica majora*, ed. H.R. Luard (London, 1876; repr. 1964), 3:115.

the bridge. Moreover, Brother Pierre Raymbert requested the restitution to the "fratres domus operis pontis" of the properties ("if they can be found"), which had been detached by Bishop Zoën to endow the priory.[36]

The actions of Bishop Zoën resulted in the loss by the brothers of their power over the bridge property, for the town authorities transferred this function to two rectors. An extant Provençal version of the life of St. Bénézet must have been written about this time, for it states that the translation was made at the request of Brother Raimon Penhere, "fraire et donat de l'obre del pont de Sant Benezet, et nos, rectors del did pont."[37] The unflattering light in which the legend places the bishop (which surprised Raynaud)[38] may be explained by the indignation of the rectors at the actions of Bishop Zoën and the bitterness of the struggle between the town and bishop. Probably the number and enthusiasm of the brethren had declined, for in 1278 the consuls commissioned persons other than the "fratres operis pontis" to collect alms for the bridge.[39] Furthermore, in 1281 there were requested and granted the first papal indulgences to those contributing to the fabric.[40]

As the order of the brothers of the bridge declined, at the end of the thirteenth century, one finds the first mention of a hospital in connection with the Pont-Saint-Bénézet. It is likely that earlier there had been a hospital (probably an overnight shelter),[41] but it is apparent that if there was a hospice, it must have been an insignificant part of the "opus pontis Rhodani." Following the removal from the brotherhood of chief responsibility for the financial affairs of the bridge, the hospital became of major concern to the brothers. During the course of the fourteenth century the hospital continued to be served by oblate brothers and sisters. Finally, after the second siege of Avignon in 1410–1411, there were no brothers or sisters to serve in the hospital.[42] Thus came to an end quietly the brotherhood of the bridge at Avignon, originally responsible for the marvel of its time and region, the mighty stone bridge across the Rhone.

The history of the brothers of the bridge across the Rhone at Lyons (see

36. Pansier, "Histoire," pp. 61–62.
37. Pansier, "Histoire," p. 21.
38. T. Raynaud, "Sanctus Ioannes Benedictus pastor et pontifex Avenione," *Opera omnia* (Lyons, 1665), 8:157.
39. Pansier, "Histoire," pp. 58–59.
40. François, marquis de Ripert-Monclar, *Bullaire des indulgences concédées avant 1431 à l'oeuvre du pont d'Avignon par les souverains pontifes* (Monaco, 1912).
41. P. Pansier, "Les Anciens Hôpitaux d'Avignon," *Annales Avignon* 15 (1929), 6. The first mention of the hospital of the bridge occurs in 1288.
42. Pansier, "Histoire," p. 33. However, there continued to be a Hôpital-Saint-Bénézet.

Fig. 15) bears a certain resemblance to that of the *fratres pontis* at Avignon. The name of St. Bénézet has been attached to each, but at Lyons in the last half of the thirteenth century it was appropriated without any basis by the "fratres pontem super Rodanum . . . peragentes"—brothers completing the bridge over the Rhone[43]—in order to excite the faithful to make donations to the structure. As at Avignon the original impulse for the construction of the bridge over the Rhone did not come from the brothers of the bridge, so at Lyons it was not they who initiated the building of the structure. The earliest extant document on this bridge, the Pont-de-la-Guillotière, concerns the cession of two plots of land by the abbot and monks of Ainay about 1180 or 1182 "on the petition of our citizens, who have under their care the fabric of the bridge."[44] Not until 1184 or 1185, two to five years after the first evidence of purchase of land for the bridge by the citizens of Lyons, do we meet with a mention of the *fratres pontis.* At that time a bull of Lucius II stated that the structure was being built by a Brother Stephen with whom other brothers were associated.[45] The document urges the faithful to give alms to the bridge and to visit the chapel. The act concludes by permitting the blessing of a cemetery, burial in which is to be restricted to the brothers, their servants (familia), and pilgrims. The insistence of the *fratres pontis* that they must have a chapel was a bone of contention between them and the local clergy.[46]

Both at Avignon and at Lyons the brothers acted as corporations to receive and administer the property of the bridge. Both groups seem to have consisted primarily of laymen, the members at Lyons being so described in a document of 1334.[47] Nevertheless, there were priests among the brothers, for in 1185 the archbishop of Lyons reserved the right to choose the priest to serve in the chapel, whether from among the brethren or elsewhere, and in 1225 the rector was Gualterius, archpriest of Mayzieux.[48] The brotherhoods at Avignon and at Lyons each acquired a chapel and a cemetery about the

43. F. Lefort, "Histoire d'un manuscrit du XIIIe siècle relatif à la contruction des premiers ponts sur la Rhône," *Trav. Acad. Reims* 76 (1884–1885), 208.
44. M.–C. Guigue, "Notre-Dame de Lyon," *Mém. Lyon* (1874–1875), p. 201.
45. M.–C. and George Guigue, "Origine du pont de la Guillotière à Lyon," *Bibliothèque historique du Lyonnais*, 1 (Lyons, 1886–1888), 129.
46. Guigue, "Notre-Dame de Lyon," pp. 204, 205, n. 1, 223. M.–C. Guigue, ed., *Obituarium Lugdunensis ecclesiae, Nécrologie* (Lyons, 1867), p. 178.
47. M.–C. Guigue, *Cartulaire municipale de la ville de Lyon* (Lyons, 1876), pp. 169 ff.
48. Guigue, ed., *Obituarium*, p. 198. Le Comte de Charpin-Feugerolles and M.–C. Guigue, eds., *Grande Cartulaire de l'abbaye d'Ainay, suivi d'un autre cartulaire redigé en 1286* (Lyons, 1876), 1:145.

same time, and at each the hospital was a very small part of the *opus pontis.*

On the other hand, there were certain important differences between the two brotherhoods. At Avignon our evidence shows the brothers to be involved in fund-raising, and in buying and administering properties, and no documents indicate that they were engaged in building the bridge. On the contrary, at Lyons the brothers were in charge of construction. It is to the *fratres pontis* that the abbey of Ainay gives the right to build certain structures; and the abbey requires the *fratres* to protect its islands and river-bank properties from the strong current (refluxus) [49] created by placing in the river piers for the Pont-de-la-Guillotière. However, evidence that the brothers actually worked with their hands is lacking.

Unlike the *fratres operis pontis sancti Benedicti* at Avignon, the brothers of the bridge at Lyons seem to have been little concerned with the collection of alms. In 1190 a document of Richard the Lion-Hearted recommended to the counts and other officials of his realm the "fratres et nuntios de ponte qui est Lugduni," who were soliciting alms for the Pont-de-la-Guillotière, but in 1209 a bull of Innocent III speaks of indulgence money to be collected by the messengers of the preceptor and brothers of the bridge.[50] Still later, in 1254, a bull of Innocent IV stated that alms would be solicited by the rectors and brothers of the Holy Spirit for the bridge at Lyons.[51] The financing of the Pont-de-la-Guillotière under the brothers of the bridge seems altogether to have been less effective than that of the Pont-Saint-Bénézet at Avignon. For one thing, at Lyons the first appeal to the pope for a bull to encourage the donation of alms was made about 1184, about the time of the start of the bridge, whereas the first indulgences granted to the Pont-Saint-Bénézet at Avignon were made only after the bridge had been in existence a hundred years. Furthermore, the Pont-Saint-Bénézet originally was erected in stone in the course of eleven years, whereas the Pont-de-la Guillotière was mostly of wood during the whole of the period that it was under the care of the brothers of the bridge.

There were other divergences between the brothers of the bridge at Avignon and at Lyons. For example, at Avignon there were both brothers and sisters; at Lyons, as far as is known, only brothers. Moreover, the formulas in the documents are different. At Avignon the head of the organization was

49. Guigue, "Notre-Dame de Lyon," p. 194, n. 1.
50. C.F. Menestrier, *Histoire civile ou consulaire de la ville de Lyon* (Lyons, 1696), Preuves, p. xxxi. Guigue, "Notre-Dame de Lyon," p. 220, n. 4.
51. Guigue, "Notre-Dame de Lyon," p. 217.

called prior, and in his absence a commander represented him; at Lyons the head was referred to as rector or preceptor. At both places the members were called "fratres pontis," but at Avignon often "fratres domus operis pontis," while at Lyons the house is rarely mentioned. Such differences render it improbable that there was any overlapping of personnel or jurisdiction between Avignon and Lyons. One may safely dismiss the conjectured that the brothers of the bridge at Lyons constituted a branch of those at Avignon, or that Brother Stephen, mentioned in the bull of Lucius III about 1184–1185, was the same as Etienne de Misendia or Manebra, prior of the house of the fabric of the bridge over the Rhone at Avignon from about 1199 to about 1213.[52]

Our information on the end of the society of the brothers of the bridge at Lyons comes from a casual reference in a document of 1334, which states that Archbishop Pierre de Savoie, "moved by just and reasonable motives," took away from laymen the administration of the chapels, hospital, the house of the bridge, the bridge itself, and property belonging to it and turned them over to the monks of the abbey of Hautecombe of the order of Cîteaux.[53] Apparently this event occurred about 1308. So ended, in about 1308, a regime which had lasted approximately a century and a quarter.

The Pont-Saint-Esprit (see Fig. 10), too, had its brothers of the bridge; and here, too, the initiative for undertaking its construction came not from them but from the citizens of Saint-Saturnin-du-Port. The *fratres pontis* are first mentioned in the extant documents only in 1277, a dozen years after the commencement of the bridge.[54] Like the *fratres pontis* at Avignon, the brothers and sisters at Pont-Saint-Esprit were described as "donati." However, they had very much less responsibility or independence than those at Avignon or at Lyons. At Avignon in the beginning the brethren had elected their own head, but at Pont-Saint-Esprit they owed obedience to the prior of the convent of St. Pierre, lord of the town, and to the rectors of the fabric. The latter were chosen by the prior from a list supplied by the townspeople. Also, in 1281 the brothers and sisters of the Pont-Saint-Esprit were under stringent regulations as to diet. In addition, they were not to eat or drink outside their own houses when in Saint-Saturnin or at the other end of the bridge.[55] The document of 1281 refers to the brothers as collectors, and this is the only

52. Guigue, "Origine," p. 130.
53. Guigue, *Cartulaire municipale*, pp. 196 ff.
54. L. Bruguier-Roure, ed., *Cartulaire de l'oeuvre des église, maison, pont et hôpitaux du Saint-Esprit (1265–1791), Mém. Acad. Nîmes,* Ser. 7, vol. 12 (1889), Annexe, p. 16.
55. Ibid., pp. 22–25.

information we have on their functions. They were not, however, the only persons soliciting alms for the bridge. Each collector, whether a brother or not, was required to operate only in his assigned diocese on pain of forfeiting his share of the alms. The same penalty was applied, if, on his return from an expedition, he failed immediately to place his receipts on the altar of the Holy Spirit, so he and the rectors could count and divide them.

It will be noted that the brothers and sisters of the Pont-Saint-Esprit sixteen years after the placing of the cornerstone are still not living communally, that they are individually paid, and that they have their own houses. This would seem to imply that they were local people.

There is no information as to whether the brothers and sisters of the Pont-Saint-Esprit performed manual labor in connection with construction. Unlike the *fratres pontis* of Lyons, they were not in charge of building the bridge, and unlike the situation at Avignon their names rarely appear in legal documents concerning the purchase of property for the fabric. Like the brothers and sisters of the Pont-Saint-Bénézet, those of the Pont-Saint-Esprit, after devoting their early stages to collecting funds, in their later period were chiefly occupied in serving in a hospital (authorized in 1308 by Philip IV). There they cared for the poor and indigent of the region.[56] In the fifteenth century they were living communally; some were priests serving the chapel, few collected funds. The right to solicit alms for the *opus* of the bridge, its hospital, and chapel was farmed, sometimes to one or another of the brothers.[57] At the Pont-Saint-Esprit, too, the society of the brothers of the bridge was *sui generis*, and there is no evidence that there was any connection with the *fratres pontis* at Avignon or Lyons.

Besides the three great bridges over the Rhone, other and smaller spans have been attributed to bridgebuilding brothers, as for example, those at Saint-Nicolas-de-Campagnac (before 1261),[58] at La Sône (Isère) (1323),[59]

56. Ibid., p. 50.

57. Ibid., p. 192; vol. 14 (1891), Annexe, pp. 275, 278 ff.

58. E. Germer-Durand, "Le Prieuré et pont de Saint-Nicholas de Compagnac, Fragment d'histoire locale," *Mém. Acad. Gard* 27 (1863), 137, 163, 185. Saint-Nicolas-de-Campagnac (Gard), com. Saint-Anastasie. Germer-Durand assumes that since in 1561 there was a confraternity of Blauzac, this organization must have existed in the thirteenth century, was of course a bridgebuilding brotherhood, and indubitably constructed the Pont-Saint-Nicolas.

59. Cant. and arr. Saint-Marcellin. In 1323 the priory of La Sône promised to maintain the bridge, chapel, and hospital from their revenues and to pay the regent of Dauphiny for the privilege. J.P. Moret de Bourchenu, marquis de Valbonnais, *Histoire de Dauphiné et des princes qui ont porté le nom de Dauphins* (Geneva, 1722),

and at Bonpas. In the first two cases nineteenth-century authors, despite a lack of evidence, have ascribed the construction of these spans to *Frères pontifes* because of an unfounded conviction that bridges of the period ordinarily were constructed by bridgebuilding brothers.

The question of Bonpas (Vaucluse) is rather different.[60] Sweeping claims have been put forward for Bonpas as the mother house of the *Frères pontifes*, from which other branches of the order were established, and this despite the fact that Bonpas displayed few of the characteristics typical of the societies of brothers of the bridge, whether at Avignon, Lyons, or Pont-Saint-Esprit. The case for considering the brothers of Bonpas to be *Frères pontifes* rests on two points: one, that some of the acts of the period refer to them as "fratres domus pontis Bonipasus"; and two, that they actually did build a bridge. The term "brothers of the house of the bridge," as the brothers of Bonpas are called when the bridge is mentioned at all, implies merely a geographical location, not an active interest in the affairs of the bridge. Members of the bridgebuilding brotherhoods at Avignon, Lyons, and Pont-Saint-Esprit ordinarily were called "brothers of the bridge" or "brothers of the house of the fabric of the bridge."

There are other points which invalidate the claim of the brothers of Bonpas to be considered a bridgebuilding brotherhood. At Avignon, Lyons, and Pont-Saint-Esprit the society was organized for the express purpose of promoting construction and maintenance of the structure after the bridge had been begun, and none of these brotherhoods outlasted its own span. On the contrary, the brothers of Bonpas had been in existence some fifty years before they built their bridge, and they survived it by some forty years. At Avignon, Lyons, and Pont-Saint-Esprit the *fratres pontis* (with the exception of an occasional priest) were laymen, but the brothers of Bonpas were monks affiliated part of the time with the Benedictines and part of the time with the Augustinians, and they fall into the category, not of brothers of the bridge, but of those numerous monasteries which at one time or another during their existence built a bridge.

The idea of bridgebuilding as a work of charity developed in the same society which produced the Peace of God movement to provide law and

---

1:286–287: Preuves, pp. 197–198. A. Champollion-Figeac, *Droits et usages concernant les travaux de construction, publics ou privées sous la troisième race des rois* (Paris, 1860), pp. 138, 140, states that this man was a Frère pontif, but on what grounds it is hard to see. See also Collin, *Le Pont*, p. 375.

60. On Bonpas and on the whole question of the bridgebuilding brotherhoods see my "The Bridgebuilding Brotherhoods," *Speculum* 39 (1964), 635–650.

order. As individuals were to be induced by moral pressure to respect the lives and property of clergy and laymen, so the hope of a heavenly reward was held out to those who would build bridges to save lives in flood time. As it had been found useful to constitute a church a corporation to receive and administer endowments, so a bridge could be treated as an *opus* and be maintained by income from its property. As the indigent were cared for in hospitals served by brotherhoods, so in the late twelfth and during much of the thirteenth century societies of the brothers of the bridge functioned as fund-raisers or administered property or oversaw construction at three great bridges in southeastern France–those at Avignon, Lyons, and Pont-Saint-Esprit. The success of the societies of brothers of the bridge depended on the enthusiastic support of their fellow townsmen and neighbors, and when they gradually ceased to receive this, the brothers of the bridge fell into decline.

4

The Bridge as a Pious Work

The Institutionalization of an Idea

AT THE TURN of the thirteenth century numerous bridges thrown across the great rivers of southern and central France—the Loire, Garonne and Rhone—bore witness to the efficacy of the idea of the bridge as private property financed by donations for the benefit of the public. Charity had built and endowed bridges whose properties were now to be administered. At this time the limitations of the idea of the bridge as a pious work became more and more apparent. For one thing, the demand for new bridges was insatiable, and charity was unable to provide them in adequate numbers. For another, all along there had been a substantial current of opinion dissenting from the idea of the bridge as a pious work, and income intended for bridge maintenance was often appropriated for other purposes.

As the concept of the bridge as an object of charity ceased to be an excitingly novel idea, its relationship with religion became institutionalized. The spiritual reward for bridgebuilding was formalized in indulgences covering a certain number of days in purgatory, and the religious aspect of bridges was made visible by providing them with chapels. The Pont-de-la-Guillotière was the object of some of the earliest indulgences (ca. 1184), and the bishop of Rodez in 1222 issued one for the bridge at Cajarc on the Lot,[1] but the proliferation of indulgences for bridges developed only after the middle of the thirteenth century. At this time the first indulgences were requested and received for repairs to the Pont-Saint-Bénézet, originally built by donations but without the aid of indulgences.

A very explicit statement of bridge construction as a pious work is contained in a bull of Pope Clement IV dated 1267. The document allowed a forty-day indulgence of penalties in purgatory to those contributing to a bridge over the Lève in the territory of Montpellier. The bull pointed out that

1. D. Rey, *Etudes archéologiques sur le vieux Millau*, 2: *Le Pont Vieux* (Millau, 1923), p. 17.

when we all stand before the judgment seat, we shall collect the fruits of the seeds we have sown on earth, and that we should perform deeds of mercy. The document recommended to the faithful a certain bridge, exceedingly necessary, which could not be finished unless a helping hand were extended.[2] It was also in the thirteenth century that legacies to bridges became commonplace. At Terrason (Dordogne) in 1233 the sum of 5 *sous* was left to the bridge, and the earliest known reference to the bridge over the Rhone at Vienne was in connection with a legacy to it in 1249.[3] Similarly, in thirteenth-century Toulouse the clergy encouraged legacies to the church, which had not been the case in the twelfth century.[4] There were frequent bequests to bridges.

In the Middle Ages the hope of a spiritual reward in the afterlife was a powerful motivating force behind gifts to endow churches, to build bridges, and to relieve the indigent and ill. It has been well said that in the sixth century the object in founding a hospital was not a question of public health but of charity.[5] It is in part because of the connection with charity and the church that chapels and hospitals were associated with bridges. Since bridges, like convents, were pious works, it was natural to think of them, too, as having the duty of hospitality and of relieving the poor. Moreover, at a time when virtually every organization, castle, and hospital had a chapel, it was peculiarly appropriate that bridges should have their own chapels to ensure divine protection for the structure and to encourage almsgiving. Documents granting indulgences sometimes specified that to obtain spiritual benefits the faithful must visit the bridge chapel, confess their sins, and make a donation to the bridge. This was true at the Pont-de-la-Guillotière at Lyons.[6] There the remission of penalties could be obtained once a year.

The vicinity of a bridge seemed to be an attractive location for hospitals and chapels, even if there were no other connection. Hospital builders preferred a site on the bank of a river for sanitary reasons, and in the special case of the flood plain of the Loire the hillock at the end of the bridge in many

2. Martène and Durand, *Thesaurus novus anecdotorum*, 2:461.
3. F. de Verneilh, "Architecture civile du moyen âge," *Annales arch.* 16 (1856), 296. C.U.J. Chevalier, ed., *Choix de documents historiques inédits sur le Dauphiné* (Montbéliard, 1874), p. 3. A will of March 1249 left 10 *sous Viennois* to the *opus pontis Rodani* at Vienne.
4. J.H. Mundy, "Charity and Social Work in Toulouse, 1100–1250," *Traditio* 22 (1966), 208.
5. J. Imbert, *Histoire des hôpitaux français* (Paris, 1947), p. 37.
6. M.–C. and G. Guigue, "Origine du Pont-de-la-Guillotière," *Bibl. Lyonnais* 1 (1886–1888), 129 ff.

cases was occupied by a chapel or hospital.[7] Elsewhere, the hospital might be at the end of a bridge in a suburb, as in the case of the hospital of Saint-Nicolas-en-Neuborg across from Metz, or it might be next to the bridge, as was the hospital of Saint-Jacques next to the Pont-de-la-Daurade in Toulouse.[8]

Evidence for a connection between hospitals and bridges is somewhat earlier than for chapels and bridges. In a document of 1166 the bishop of Avignon turned over to masters and hospitalers or builders of the bridge at Bonpas one half the tolls.[9] At a number of places, as at Orleans, La Sône (Isère), and Romans, the bridge, its hospital, and its chapel were all under the same management. At Romans all three belonged in the thirteenth and fourteenth centuries to the chapter of St. Bernard. Rebuilt by Jean de Bernin, abbot and also archbishop of Vienne, by 1252, the bridge was supported by three stone piers. The pier nearest the town carried on one side of the roadbed a chapel, dedicated to the Virgin Mary, and on the other a lying-in hospital.[10]

Where an *opus pontis* combined a hospital and a bridge, sometimes the interests of the one predominated, sometimes of the other. In the case of the Pont-de-la-Guillotière the primary object was the bridge across the Rhone, and the hospital, in existence by 1184 or 1185, was insignificant; nor was the almonry, under the management of the brothers from 1226, a major object of their concern.[11] On the other hand, at Martigues (Bouches-du-Rhône) the bridge was subordinate to the hospital. This bridge, leading to the island of Saint-Geniez, was built, together with a church, by the hospital of the order of the Holy Spirit, as we know from an act of Hugues des Baux, viscount of

7. R. Dion, *Le Val de Loire, étude de géographie régionale* (Tours, 1934), p. 296. There were chapels at Pontvien (Mayenne), commune Livr, in 1184 (capellam de Ponte-Viviani); at Pontpierre (Mayenne) in 1228; at Pont-Pierre (Yonne) before 1290; and at the Pont-des-Natiaux (Yonne) in 1296. See appropriate topographical dictionaries.
8. L. Larchey, "Notice historique sur l'hôpital Saint-Nicolas de Metz," *Mém. Acad. Metz* 34 (1852–1853), 181, 189. Mundy, "Charity," p. 224. In 1230 at Compiègne there was a legacy left to the hospital of Saint Nicolas "de ponte Compendini": E.E. Morel, ed., *Le Cartulaire de l'abbaye de St. Corneille de Compiègne*, 2 vols. (Montdidier, 1904–1909), 2:80, #CCCXCIV. At Albi in 1157 there was a hospital of Bout-du-Pont, the settlement at the end of the bridge: "Glanures historiques," *Rev. Tarn* 8 (1890–1891), 212.
9. R. Michel, "Les Constructions de Jean XXII à Bonpas," *Mélanges Ecole franç. de Rome* 31 (1911), 381.
10. U. Chevalier, "Notice historique sur le pont de Romans," *Bull. Drôme* 2 (1867), 312–313.
11. M.–C. Guigue, "Notre-Dame de Lyon," *Mém. Lyon* (1874–1875), p. 223.

Marseilles, who in 1211 took under his protection the hospital of the Holy Spirit of Marseilles, its dependent hospital of Saint-Geniez at Martigues, and the bridge of Saint-Geniez. In 1212 the bishop of Uzès, apostolic legate, exhorted the faithful to aid with their alms the *opus* of the church, bridge, and hospital of Saint-Geniez.[12]

The hospital of Saint-Nicholas-en-Neubourg at Metz was exceptional in having the responsibility for a number of bridges, collecting not only tolls but a species of tax to support them. The "warnement des morts" consisted in the payment of the best outfit of clothes of every deceased citizen of Metz for the support of the bridges.[13] This tax had originally been instituted by the bishop to pay for the construction of one of the bridges at Metz—the Middle Bridge, 1222–1223. Such a tax was a variation of the death due consisting of the bed in which the deceased died, paid at many places in France to the local hospital.[14] The hospital of Saint Nicolas was tightly controlled by the town of Metz, which seems to have considered its property at municipal disposal. In 1267 the right to collect bridge tolls from strangers and to inherit the best clothes of the deceased from the inhabitants of Metz had belonged half to the hospital of Saint-Nicolas and half to the leprosary of Saint-Ladre. In 1262 the totality of these rights was acquired by the hospital on the payment of 1100 *livres messines* and the promise that the hospital would rebuild the bridges in stone, beginning six years after the purchase

12. Archives départementales des Bouches-du-Rhône. Livre trésor de l'hopital Saint-Esprit de Marseille, 1399, fols. 57v, 58v, 59r. A. Fabre, *Histoire des hôpitaux et des institutions de bienfaisance de Marseille* (Marseille, 1854–1855), 1:44–46. At Château-Gontier the brothers of the almonry situated at one end of the bridge over the Mayenne received in 1206 certain tolls at vintage time in consideration of replacing planks of the bridge as required. Evidently they remained responsible for the bridge at least during the remainder of the Middle Ages: E. Moreau and A. G., "Une Vue du vieux pont de Laval, vers 1676, et du pont de Château-Gontier," *Bull. Mayenne*, 2e ser., 29 (1913), 240. A. Angot, *Dictionnaire historique, topographique et biographique de la Mayenne* (Laval, 1900–1910), 1:577. The bridge at Château-Gontier is believed to have been built as part of the fortifications at that place by Fulk of Anjou ca. 1007, and it is attested about 1080–1096: M. du Brossay, "Cartulaire d'Azé et du Géneteil, prieuré de l'abbaye Saint-Nicolas d'Angers," *Arch. Maine* 3 (1903), 52.
13. Before the corpse was buried, "son millour chaperon et son milour warnement de robbes" were to be given for the bridges: Larchey, "Notice," p. 181.
14. Imbert, *Histoire des hôpitaux*, pp. 297–298. The monastery of St. Pierre of Saint-Saturnin-du-Port inherited the furnishings of the room in which an inhabitant of the town died, but in the fourteenth century the community preferred to pay 30 florins to be quit of this right: L. Bruguier-Roure, ed., *Cartulaire Saint-Esprit*, *Mém. Acad. Nîmes*, Ser. 7, vol. 12 (1889), Annexe, p. 46, n. 1.

date.[15] Lawsuits by local convents were unsuccessful in procuring exceptions from the *warnement des morts.* Many neighboring villages preferred to pay this mortuary tax rather than the tolls, and in the fourteenth century members of families from these villages came annually to the hospital to swear on the altar that they were "bourgeois of the hospital" and prepared to pay the clothes of their dead as dues.[16] However, the hospital of Saint-Nicholas found the income from tolls and death dues inadequate to maintain the hospital and bridges. Only in 1336 did the hospital of Saint-Nicolas begin the promised reconstruction in stone of the Pont-des-Morts, Pont-Thiffroy, Pont-Saint-Georges, and Middle Bridge.[17] The town of Metz was obliged to come to the aid of the hospital and grant it additional sources of revenue.

The idea of a hospital as part of the *opus pontis* had appeared in the second half of the twelfth century, but it was not a particularly happy combination, and in many cases the connection was not long-lasting. At Bonpas the bridge and hospital, mentioned in 1166, had both disappeared by 1189. At Avignon the hospital of the bridge, which first comes to notice in 1288, was no longer functioning in 1410; and at Lyons the hospital with a chapel was detached from the *opus pontis* and transferred to the abbey of Chassagne in 1334. It was exceptional that as late as some time between 1344 and 1386 the hospital of Saint-Antoine at Orleans was added to the fabric of the bridge, and that the administration of the hospital of Saint-Nicolas at Metz continued to be in control of bridge construction and maintenance from the thirteenth century throughout the Middle Ages.

Chapels, unlike hospitals, continued to form part of the fabric of the bridge, and they were still being built on bridges as late as the eighteenth century. It was not uncommon to find the chapel on a cutwater. Still to be seen are those at the Pont-Saint-Bénézet (Fig. 8) and at Mailly-le-Château (Yonne). In the Middle Ages the numerous chapels connected with bridges were dedicated to a variety of saints. St. Nicholas was the patron of chapels on bridges at Avignon, the Pont-Saint-Esprit, Pont-de-la-Guillotière, and Mailly-le-Château; but at Blois the patron was St. Fiacre, at Montauban St. Catherine, at Clamecy (Nièvre) St. John, at Beaugency St. Jacques, and at Romans the Virgin.[18] Chapels dedicated to the Holy Spirit formed part of the *opus pontis*

15. Larchey, "Notice," pp. 181–183.
16. Larchey, "Notice," p. 203. The inhabitants of some villages paid an annual loaf of bread to the hospital for an exemption from tolls: J. Schneider, *Le Ville de Metz aux XIIIe et XIVe siècles* (Nancy, 1950), p. 15.
17. Larchey, "Notice," p. 182.
18. For Mailly-le-Château and Clamecy see M. Quantin, "Histoire de la rivière d'Yonne,"

at the Pont-de-la-Guillotière and at the Pont-Saint-Esprit. The chapel and hospital located on the island crossed by the bridge at Orleans were under the protection of St. Anthony. As early as the 1180s chapels formed part of the *opus pontis* at Avignon and at Lyons; and as the number of bridges increased, chapels proliferated. Moreover, some bridges had more than one chapel. In general, the earlier chapels were more likely to be built on the bank near the bridge, and the later chapels on the bridge. Thus at Lyons and at Clamecy in the twelfth century there were chapels nearby on the bank, but in the fourteenth century each of them also had one built on the bridge.

From the beginning of the idea of the bridge as a pious work there had been difficulties, sometimes resulting from a lack of enthusiasm on the part of the faithful and sometimes due to competing interests and property rights. Bridges shared with churches, convents, and hospitals the distinction of being works of charity. A Toulouse testament of 1251 summed up the matter succinctly in leaving sums to "churches and hospitals and bridges and other pious and poor places."[19] Such a syntagma falls strangely on modern ears. It appears to us to confuse spiritual needs, public health, and the administrative province of the department of bridges and highways. Even in its heyday this lack of distinction among objects of charity met with objections. All these institutions–churches, convents, hospitals, and bridges–were competing for the support of the faithful, and perhaps it is not surprising that some ecclesiastics should have been convinced that the most important function of a pious work, whether a hospital or a church, was performance of the mass.[20] Anxiety to avoid dissipation of the largesse of the faithful was important in the history of the bridge as a pious work.

The erection of a chapel on or adjacent to a bridge and dependent on the *opus pontis* sometimes met with opposition from the clergy in whose parish it was situated. Perhaps for this reason the convents of Aniane and of Saint-

---

Bull. Yonne 39 (1885), 426. For Blois see A. de Martonne, "Notice historique sur l'ancien pont de Blois et sa chapelle," *Mém. Orléanais* 6 (1863), 415. For Montauban see Le Chanoine Pottier, "Le Pont de Montauban," *Bull. Tarn-et-Garonne* 1 (1869), 34. For Beaugency see A. Collin, *Le Pont des Tourelles à Orléans (1120–1760)* (Orleans, 1895), pp. 215, 217. In 1204 a chapel dedicated to St. Nicholas was planned next to the bridge at Charenton: J. Depoin, ed., *Recueil de chartes et documents de Saint-Martin-des-Champs, monastère Parisien* (Paris, 1912–1921), 3:#616. For Romans see Chevalier, "Notice historique," p. 313.

19. Mundy, "Charity," p. 205, n. 4.

20. Fabre, *Histoire des hôpitaux*, 1:33. The provincial council of Arles complained in 1260 that in Provence a great number of charitable establishments were almost useless to the unfortunate, because clerks and laymen appropriated the revenues.

Guillem-le-Désert agreed (sometime between 1036 and 1048) that there should be no church on their new bridge over the Hérault at the Gouffre-Noir.[21] Nevertheless, toward the end of the twelfth century chapels came more and more to be associated with bridges. At Lyons in 1185 the abbot and monks of Ainay contended that the bridge chapel on the banks of the Rhone was prejudicial to the parish church of Saint Michael, and the brothers of the bridge were obliged to agree to remove the chapel within ten years.[22] The abbey in 1226 allowed the brothers to erect a new chapel to the left of the bridgehead,[23] but in the middle of the century the convent opposed a chapel newly built "in capite dicti pontis" by the brothers of the bridge as being to the prejudice of the monastery. Their objections were overuled by a secular court, which gave the chaplain, preceptor, and brothers of the house of the bridge permission to repair the chapel in stone and wood from foundation to rooftree in any way they might prefer.[24]

The brothers of the bridge at Lyons were able to defend their chapels against the local church, but the rectors of the Pont-Saint-Esprit were less successful in protecting from the monastery of St. Pierre of Saint-Saturnin funds donated to the bridge in its oratory. As the bridge neared completion, the rectors wished to establish a hospital in connection with it, and at the same time the prior and monks of St. Pierre cast longing eyes on the funds of the Pont-Saint-Esprit. Construction of a larger parish church to accommodate the influx of workmen for the bridge was a heavy burden on the convent's finances. The plan for a bridge hospital seemed to threaten the monastery's share of the alms of the faithful, especially the convent's hospital. Feelings ran so high that in 1297 the monks tore down the wall of the rectors' hospital, then under construction.[25] The rectors' suspicions of the prior's intentions were increased when in 1301 he insisted that the rectors of the bridge of the Holy Spirit should render account before him and his lieutenant alone, rather than before him and the inhabitants of Saint-Saturnin, as had formerly been the case.[26]

In 1306 only one pier and one arch of the Pont-Saint-Esprit remained to

21. C. Devic and J. Vaissète, *Histoire générale de Languedoc* (Toulouse, 1872–1893), 5:col. 393.
22. M.–C. Guigue, *Obituarium Lugdunensis ecclesiae* (Lyons, 1867), pp. 178, 198. Idem, "Notre-Dame de Lyon," pp. 203–204.
23. Guigue, "Notre-Dame de Lyon," pp. 204–205.
24. Guigue, "Notre-Dame de Lyon," p. 224.
25. Bruguier-Roure, *Cartulaire Saint-Esprit*, *Mém. Acad. Nîmes*, ser. 7, vol. 12 (1889), Annexe, p. 28
26. Ibid., pp. 31–34.

be constructed.[27] The determination of the prior to avail himself of the funds given to the *opus* of bridge, chapel, and hospital in the oratory of the Saint-Esprit was met by that of the rectors who insisted that maintenance of the bridge would be very expensive and that there was no money to spare. Guillaume de Plaizan, representing the king of France, in 1307 handed down an arbitral award in the controversy. Despite a tribute to the Pont-Saint-Esprit as a marvellously great bridge which provided a safe crossing and eliminated the dread of shipwreck,[28] the decision provided that half the collection at the oratory was to go to the convent, which, however, was to have no part in any other offerings to the bridge.[29] The rectors of the *opus pontis* were to be chosen annually from among the *probi homines* of the town of Saint-Saturnin. They were to administer the affairs of the bridge and of the hospital for the indigent and to report to the seneschal of Beaucaire (rather than to the prior and the men of Saint-Saturnin). Finally, the men of Saint-Saturnin were to give to the prior 400 *livres tournois* towards the completion of the church, a sum later reduced by one half.

The acrimonious tone of the dispute at the Pont-Saint-Esprit is a reminder that the bridge at Saint Saturnin was begun nearly ninety years later than the Pont-Saint-Bénézet. The cornerstone was laid in August 1265, and its completion required more than forty years rather than the eleven sufficient for the bridge at Avignon. In the later period there was no longer the same generosity in supporting pious works, and the claim of the bridge to charity was meeting active competition from others demanding alms. The bequests, the indulgences, and the legacies continued, but the real support for the Pont-Saint-Esprit was the donation about 1326 by King Philip VI of the *Petit Blanc*, a tax on salt passing under the bridge or sold in Saint-Saturnin. This charge enabled the bridge to be maintained toll-free and in good condition.[30]

The protection which should have been extended to all pious works failed bridges when they became involved in wars. During hostilities between the chapter of St. Bernard and the inhabitants of Romans the former built a tower on the bridge. The Romanais seized the fortification. In 1281 the bishop of Valence came, ruined the tower, turned out of doors twenty-three women in the lying-in hospital, removed the Virgin's statue from the chapel,

27. Ibid., p. 33.
28. Ibid., p. 41.
29. Ibid., p. 47.
30. Ibid., vol. 17 (1894), Annexe, p. lxii.

burned the buildings on the bridge, and took the town. The inhabitants were condemned to pay a fine to rebuild the bridge and hospital.[31]

Although the Pont-Saint-Bénézet was the most immediately successful of the great bridges built by donations, it was the one that aroused the enmity of the king of France and the greatest opposition from the clergy. During the Albigensian war Avignon sided with Raymond VII, count of Toulouse. The bridge was damaged during the three-month siege of the city by Louis VIII of France in 1226, and the terms of surrender called for its destruction and the razing of the town walls. The commune was forbidden to rebuild its fortifications or bridge for a period of five years.[32] It is to be noted that the legate, far from considering the building of the bridge a work meritorious in a spiritual sense, was treating it as part of the fortifications of a rebellious town allied with heretics. As soon as the political situation began to improve, the Avignonais took up the task of rebuilding their walls and bridge.[33] By 4 May 1234 they had made some progress, for on that date, despite the fact that more than five years had elapsed, a bull of Gregory IX ordered them immediately to demolish their fortifications and bridge.[34] Apparently the bull was ignored. A bull of 8 August 1237, in commanding the Avignonais to desist from the reconstruction of their ruined bridge and of their gates, walls, and moats, referred to such activities as "in contempt of religion."[35] Nevertheless, the bridge was rebuilt, but probably in wood rather than stone.

In what is now France it was especially in the South and along the Loire that the idea of the bridge as a pious work flourished. Almost all chapels connected with bridges were situated in these areas, and almost all legacies were devised to, and indulgences authorized for, bridges in the same regions of France. It is understandable that this should have been so because of the separate funding of bridges there. An endowed bridge was capable of receiving a legacy and of maintaining a chapel from its own funds, and the *opus pontis* does not seem to have occurred elsewhere in what is now France.

Even at the peak of enthusiasm for the idea of the bridge as a pious work, and even in that region of France where it especially flourished, there were difficulties in funding bridges through charity, as demonstrated in cases at Agen, Carcassonne, and La Rochelle. As to La Rochelle, there was obviously

31. Chevalier, "Notice historique," p. 313.
32. L.H. Labande, *Avignon au XIIIe siècle* (Paris, 1908), pp. 30–38, 48.
33. Labande, *Avignon*, p. 48.
34. L. Auvray, *Les Registres de Grégoire IX* (Paris, 1896–1955), #1912.
35. Auvray, *Registres*, #3802.

an oversimplification both in the statement of King John in 1202 that Master Isembert had built the bridge with the "alms of the faithful" and that he had donated the bridge to certain persons.[36] Evidently the brothers of the Alms House had paid Isembert for his "donation" to them. A document of 1207 shows the ownership of the bridge and its properties to be equally divided between the brothers of the Alms House and the brothers of the Temple of La Rochelle, but it is stated that the former remain deeply in debt to the late Master Isembert. Accordingly, the settlement of a quarrel between the two parties provided that the brothers of the Alms House should receive the first 60 *livres poitevins* of income from the bridge to discharge their debt to the late master of the schools. After that, all income and expenses were to be divided equally between them.[37]

Charitable resources were too limited to provide the wanted bridges. An example of the failure of the appeal to alms is provided by the history of Agen. At about the same time that the brilliantly effective fund-collecting of Saint-Bénézet and his brothers of the bridge was financing the immense stone bridge across the Rhone, the town of Agen procured from Richard the Lion-Hearted in 1189 a charter authorizing the collection of donations to build a bridge across the Garonne. Two persons to be chosen by the common council of the commune and by the *pontonnarius* (the man in charge of the bridge) were permitted to solicit funds. Passage across the structure was to be free of any toll, *pontenage*, or charge, and since donations would be needed year after year, first for construction and subsequently for maintenance, provision was made for successors to the collectors on their demise.[38] However successful the recourse to charity may have been elsewhere, at Agen it seems to have resulted in funds sufficient to erect a single pier only. Evidently the latter was of massive bulk, for about 1217 it was deflecting the river away from the city, thus removing its best defense. Accordingly, the town applied to Count Raymond of Toulouse and received permission to demolish this solitary witness to the hopes of Agen.[39]

Nearly a hundred years after the first attempts there is evidence of re-

36. V. Mortet and P. Deschamps, eds., *Recueil de textes relatifs à l'histoire de l'architecture et à la condition des architectes en France, XIIe–XIIIe siècles* (Paris, 1929), p. 194.
37. L.M. De Richemond, "Chartes de la Commanderie magistrale du Temple de la Rochelle (1139–1268)," *Arch. Saintonge* 1 (1874), 35 ff.
38. G. Tholin, "Les Ponts sur la Garonne. Extrait de l'Abregé chronologique des antiquités d'Agen, par Labrunie," *Rev. Agenais* 5 (1879), 440.
39. Tholin, "Les Ponts," p. 441, n. 2.

newed activity.[40] Now the initiative came not only from the clergy but also from the royal government. On 25 March 1283 and again on 14 November 1284 a circular from the king of England was sent by the vice-seneschal of the Agenais to order the bailiffs, provosts, and consuls of the province to bring together their people to invite them to contribute to the bridge of Agen, where work was now again being undertaken after a long lapse of time. Still another effort to finance construction by donations was contained in a letter of the bishop of Agen, dated May 1286, to the clergy of his diocese, prescribing the collection of an offering for the bridge during one month, during which time donors were to receive forty days' relaxation of ecclesiastical penalties. Also, the king of England reissued the letters of Richard I, authorizing solicitation of funds; but all was in vain. No appeals to charity were effectual in financing the bridge.

Slightly earlier than Agen's fruitless efforts to finance a bridge through charity, the town of Carcassonne erected across the Aude the stone bridge still to be seen today. Surviving documents give no hint that the construction was in any way dependent on charity. In one, Viscount Roger of Béziers regrets that he had collected tolls on the bridge, because he had formerly freely and absolutely given the men of Carcassonne license to build it.[41] In the other, dated 1184, the viscount, in consideration of an annual payment of two measures of grain, donated to the men of Carcassonne the bridge they had built, and, significantly, he included in the gift the right to collect taxes, provided all such monies were sent to the *opus pontis.*[42] The incident illus-

40. T.N. Bisson, "An Early Provincial Assembly: The General Court of Agenais in the Thirteenth Century," *Speculum* 36 (1961), 254–281. Tholin, "Les Ponts," pp. 441–442. *Inventaire-sommaire arch. com. Agen*, DD, pp. 3–4.

41. E. Baluze, *Histoire genéalogique de la maison d'Auvergne* (Paris, 1708), vol. 2, Preuves, p. 500: "Quod in transitu pontis Carcassonae accipiebam, injuste accipiebam, quia olim dederam licentiam faciendi pontem hominibus Carcassonae libere et absolute; & illud idem confirmo & coroboro in perpetuum per me & per omnes successores meos." See also [J.–P.] Cros-Mayrevieille, *Les Monuments de Carcassonne* (Paris, 1850), p. 186, and pp. 17, 63, 93–96, 187, n. 3.

42. A.J. Mahul, ed., *Cartulaire et archives des communes de l'ancien diocèse et de l'arrondissement administratif de Carcassonne*, 6 vols. in 7 (Paris, 1857–1886), 5:313: "Dono sine fine, vobis omnibus hominibus ville Carcassone, praesentibus et futuris, scilicet illum pontem ville Carcassonne, situm super Atacem, cum omnibus sibi pertinentibus et conquerimentis que ibi et acaptes facere poteritis, et ut licentiam juxta voluntatem vestram habeatis et ad requirendi, et acaptandi, atque collectam faciendi ubicumque et in quibuscumque hominibus volueritis ad opus pontis illius, et vos illas acaptes, et omnem illam fideliter in opere pontis mittatis." Devic and Vaissète, *Histoire*, 8:col. 374. See also A. Champollion-Figeac, *Droits et usage concernant les travaux de construction* (Paris, 1860), p. 131.

trates the future course of the history of medieval bridges in France, in that the bridge was built by townsmen and maintained by tolls and taxes. In this, Carcassonne was ahead of its time, perhaps because of a superior degree of urbanization in the South or perhaps because of a persistence of the ideal of public utility.

The practical problems of the bridge as a pious work—conflicts of interest or inadequacy of alms—made no difference to the theory of the bridge as an object of charity. The ideal of the bridge as a pious work persisted throughout the remainder of the Middle Ages. In the fourteenth and fifteenth centuries indulgences for bridges continued to be granted; more chapels were constructed on spans, and the *opus* of many a bridge continued to send out collectors. However, the institutionalization of the idea of the bridge as a pious work was no substitute for the initial enthusiasm. As bridges, once built, were found to require routine maintenance and emergency replacement after catastrophic floods, the demands of bridges began to appear endless to the weary faithful. At the same time a decline in giving threatened churches and hospitals. The solution to the problem of bridge construction and maintenance appeared to be tolls and taxes.

5

Bridges for Private Advantage

THE ROMANTIC FASCINATION of nineteenth-century writers with the medieval view of bridge construction as a pious work has obscured the fact that in France this was essentially an idea of the South and Center. It hardly occurred north of the Loire. Even at the time of the flowering of the concept it can be shown that in southern and central France many bridges were built without reference to the notion. Spiritually, at this time the bridge was a good work built to preserve men from drowning. In practical terms, a bridge might determine the course of a military campaign, alter trade routes, support an abbey, destroy the profits of a ferry owner, or change the pattern of giving of the faithful. Perhaps it is not surprising that even while potential donors were being admonished to save lives in flood time or to improve a region by building bridges, many churches, convents, and lay lords were treating bridges as private property. There are numerous examples of cases in which the owner of a bridge viewed it in terms of his own defense or profit, seemingly without consideration of the general benefit. Despite its establishment in the public interest, the *opus pontis* was a private corporation, although ordinarily a local bishop or sometimes a neighboring town had a measure of control over it.

Frequently in the eleventh, twelfth, and early thirteenth centuries possession of a bridge was disassociated from any public responsibility. Formulas for conveying the ownership of a bridge seem to convey full ownership, as for example at the Pont-de-la-Daurade and the Pont-Saint-Bénézet.[1] A royal document confirming the canons of Saint-Corneille of Compiègne in possession of their bridge in 1112 is very explicit:

> Concedimus eidem sancte Compendiensi ecclesie ad clericis ibidem Deo servientibus, quatinus predictum pontem restituant teneant ac perpetuo

1. G. Catel, *Histoire des comtes de Tolose* (Toulouse, 1624), p. 196. P. Pansier, "Histoire de l'ordre des frères du pont," *Annales Avignon* 7 (1920–1921), 15.

absque ulla contradictione possideant, habeantque licentiam ac plenissimam potestatem faciendi eum et meliorandi.[2]

It was the men upon whom the powers of the central government devolved who built almost all the bridges in the eleventh century and many in the twelfth in France. Lay and ecclesiastical lords were responsible for many spans. It is not accidental that of three towns named "Bishop's Bridge," one (Pont-Evêque [Isère]) dates from the eleventh century and two (Pont-l'Evêque [Calvados], Pont-l'Evêque [Finistère]) from the twelfth. Between 1024 and 1093 a bishop built a bridge over the Vire near Coutances (Manche).[3] In the late twelfth century Maurice de Sully, bishop of Paris, who began to build the cathedral of Notre Dame, also constructed two stone bridges in 1175, one on the Seine and one on the Marne.[4] Before 1143–1145 the monks of Saint-Léonor were responsible for the bridge at Beaumont-sur-Oise, and in 1205 the bishop of Soissons and the cathedral chapter of Châlons-sur-Marne decided that half the alms donated because of the relic of Saint Stephen should go for building the cathedral and the other half to the construction of the city bridge. The latter was completed in 1217.[5]

Where the local counts or dukes possessed well-organized fiefs in the eleventh and twelfth centuries, a correspondingly greater number of bridges was built in that period, in some regions as many as in the thirteenth century. Along the Loire and its tributaries (in the counties of Maine, Poitou, Anjou, Blois, and Touraine) 22 bridges first come to notice in the eleventh century, 20 in the twelfth, 20 in the thirteenth, only 5 in the fourteenth, and 23 in the fifteenth. Thus Fulk Nerra, count of Anjou, constructed two stone bridges in 1028, one at Angers and one at Mayenne, and it was local lords who built bridges at Amboise in 1110 and at Chalonnes-sur-Loire[6] in the

2. E.E. Morel, ed., *Cartulaire de Saint-Corneille de Compiègne* (Montdidier, 1909), 1:64.
3. V. Mortet, ed., *Recueil de textes relatifs à l'histoire de l'architecture, XIe–XIIe siècles* (Paris, 1911), pp. 70, 73. See also topographical dictionaries.
4. "Ex chronologia Roberti Monachi Sancti Mariani Altissiodorensis," *Historiens de France*, 12:298.
5. J. Depoin, ed., *Recueil . . . Saint-Martin-des-Champs* (Paris, 1912–1921), 2:#285 ter. For Chalons-sur-Marne see V. Mortet and P. Deschamps, eds., *Recueil de textes relatifs à l'histoire de l'architecture, XIIe–XIIIe siècles* (Paris, 1929), p. 194.
6. For bridges at Angers, Mayenne, and Chalonnes-sur-Loire, see J.M. Bienvenue, "Recherches sur les péages angevins au XIe et XIIe siècles," *Le Moyen-Age* 63 (1957), 213, 218. For Amboise see "Ex gestis ambasiensum dominorum," *Historiens de France*, 12:510 (in 1110). For Chalonnes-sur-Loire see also C. Urseau, ed., *Cartulaire noire de la cathédrale d'Angers* (Angers, 1908), p. 294.

middle of the twelfth century. In Normandy, where the duchy was exceptionally tightly controlled at an early date, there were already as many bridges being built in the eleventh century (11) as in the twelfth (10), thirteenth (10), and fourteenth (9). The bridge across the Seine at Rouen is attested by 1027,[7] and in the twelfth century it was the dukes of Normandy who maintained it. After the structure was burned in 1136, it was rebuilt by Geoffrey Plantagenet, count of Anjou and duke of Normandy. Later, in 1151–1167, it was reconstructed by the Empress Matilda, heiress of Normandy.[8]

In contrast, in the Ile-de-France and the remainder of the area where the kings of France first organized their domain in the course of the twelfth century, 8 bridges come to notice in the eleventh century, 17 in the twelfth, and only in the thirteenth does the number rise to 34; but the number sinks to 11 in the next century, during the Hundred Years' War.

In Champagne, where the efforts of the counts to encourage their fairs are well known, it is probable that they also promoted bridge construction. The efflorescence of the fairs occurred in the late twelfth and in the thirteenth centuries, and these two centuries correspond with the great era of bridge construction in the county. In Champagne 27 bridges come to notice in the twelfth century and 26 in the thirteenth, whereas there are only 6 new ones attested in the fourteenth.

It may be conjectured that reasons of public utility accounted for the construction of some at least of the above-mentioned bridges in Normandy, Maine, Poitou, Anjou, Blois, Touraine, Champagne, and Ile-de-France, but it is much easier to demonstrate that motives of defense, profit, or the builders' convenience played a powerful role in deciding where bridges should be built and whether they should be built. A significant determinant was military advantage. In the Middle Ages (unlike the modern period) the defense in wartime had an advantage over the offense, a fact which accounts for the myriad ruins of castles dotting the landscape of Europe. Drawbridges were a usual part of fortifications, but in addition bridges in the vicinity of a castle were frequently essential to the defense. About 1007 Fulk of Anjou built the bridge at Château-Gontier as part of the fortifications of his castle.[9] About

7. A. Deville, "Recherches sur l'ancien pont de Rouen lues à la séance du 10 décembre 1830," *Précis analytique. Travaux Acad. Rouen* (1831), p. 169.
8. "Ex Roberto abbatis de Monte. Appendice ad Sigebertum ad calcem operum Guiberti Novigenti abbatis," *Historiens de France*, 13:290, 311.
9. A. Angot, *Dictionnaire historique, topographique et biographique de la Mayenne*, 4 vols. (Laval, 1900–1910), 2:577.

1025 the count of Sens rebuilt bridges across the Seine and Yonne at Montereau-Faut-Yonne to connect his fortress, situated on a tongue of land, with the opposite banks.[10] At Pont-de-l'Arche the bridge over the Seine was part of the citadel of the dukes of Normandy as early as ca. 1020 and it still continued such at the end of the twelfth century.[11] At Pierre-Châtel (Ain) the count of Savoy, who had a fortress there, rebuilt the bridge after the flood of 1226, and the bishop of Auxerre in the thirteenth century constructed a stone bridge across the Loire at Jargeau, where he had a castle.[12] The efforts of Richard the Lion-Hearted to fortify his duchy against his overlord, Philip Augustus, king of France, were responsible for the construction of a number of bridges. Richard in 1197 built four spans at Les Andelys (Eure) in connection with the Château-Gaillard.[13] However, after the English had been expelled from Normandy, Maine, and Anjou, and order was established in most of the kingdom under Louis IX, for many years there was little occasion for bridge construction in connection with fortifications.

It can be shown that convents derived advantages from bridge construction facilitating communication with the outside world. Nevertheless, there is no reason to adopt the excessively pessimistic view of F. de Verneilh as to bridge construction for purely religious ends on the part of convents.[14] The motives of Abbot Peter of Andres in building the bridge at Ausque near Thérouanne (Pas-de-Calais) probably were in fact, as the chronicler described

10. E.E. Viollet-le-Duc, *Dictionnaire raisonné de l'architecture française*, 7 (Paris, 1864), 238.
11. *Dic. top. Eure*, s.v. L. Delisle, L. Passy, eds., *Mémoires et notes de Auguste Le Prevost* (Evreux, 1864), 2:573. Bridges constructed in connection with castles are attested at Châteauneuf-sur-Sarthe (Urseau, *Cartulaire noir d'Angers*, p. 285); at Gournay-sur-Marne (Mortet, *Recueil*, p. 335); at Coulanges-sur-Yonne (Yonne) (M. Quantin, "Histoire de la rivière d'Yonne," *Bull. Yonne* 39 [1885], 425); at Pont-du-Château (Puy-de-Dôme) (Suger, *Oeuvres complètes*, ed. Lecoy de la Marche [Paris, 1867], p. 123). *Dic. top. Aube*, p. 126. In 1212 fortifications at Melun included two bridges: Mortet and Deschamps, *Recueil*, p. 215.
12. For Pierre-Châtel (Ain), see *Dic top. Ain*, s.v. The bishop of Auxerre built stone bridges across the Loire at Jargeau and at Meung: Mortet and Deschamps, *Recueil*, p. 202.
13. T. Bonnin, ed., *Cartulaire de Louviers: documents historiques originaux du Xe au XVIIIe siècle* (Evreux, 1870–1883), 1:101n. In 1195 Richard had rebuilt the ogival arch of the bridge next to the town of Pont-de-l'Arche, and at Portejoie (Eure) he was responsible for a wooden bridge with a small tower: Delisle and Passy, *Mémoires*, 2:574; Mortet and Deschamps, *Recueil*, p. 174, n. 4. For Gisors see Mortet and Deschamps, *Recueil*, pp. 107–108. See also p. 118 for another bridge in connection with a castle (1170–1191).
14. "Architecture civile du moyen-âge," *Annales arch.* 16 (1856), 296.

them, "the love of God and the repose of his poor," including pilgrims to the monastery.[15] It was afterwards that the chronicler wryly notes that the results of Abbot Peter's philanthropic project were not altogether happy. The parishioners were ungrateful. There were complaints from ferrymen deprived of income, stones were removed from the parapets by washerwomen seeking to ease their labors, and finally, since Abbot Peter was known to have first built the bridge, its maintenance became a charge upon the abbey under his successors.

Some bridges were constructed so that properties could be exploited conveniently, as when in 1095 the canons of Saint Ambroix secured from the viscount of Bourges permission to rebuild the bridge and road to their vineyard, or the abbey of Toussaint-en-l'Ile was allowed by the bishop of Châlons-sur-Marne (in the years 1164–1191) to build new houses throughout the island and a bridge by which one could freely cross to the city.[16]

Profit might be the chief interest of the builder or owner or the only consideration of a recipient of tolls who acknowledged no obligation to maintain the bridge. Bridges as a source of revenue may be illustrated by that at Juvardeil-sur-Sarthe (Maine-et-Loire). In 1075 the lord of Juvardeil and the monks of the priory of that place agreed to finance jointly the construction of a bridge and to share equally the income and the cost of repairs.[17] The monks of Saint-Leónor, who built the bridge of Beaumont-sur-Oise, were guaranteed by the count of that place in 1143–1145 a perpetual annual rent of 100 *sous parisiensis* and 10 measures of salt from the bridge toll.[18] An enterprising prior of the convent of Gournay-sur-Marne obtained from the count of Meulan in 1147–1154 permission to construct a stone bridge across the Marne with mills in the arches, and a rent of 100 *s. p.* yearly to be paid, not from the bridge tolls, but from those of the castle of Gournay. This agreement was never ratified by the monks, and in the twelfth century there seems to have been a wooden bridge at Gournay.[19] Nevertheless, the profit motive for bridge construction on the part of the prior would seem to be entirely clear.

In the early twelfth century ownership of a bridge could be a perquisite

15. Mortet, *Recueil*, p. 392.
16. For Saint-Ambroix see Mortet, *Recueil*, p. 103, n. 2. For Châlons-sur-Marne see Mortet and Deschamps, *Recueil*, pp. 114–115.
17. P. Marchegay, "Chartes angevines des onzième et douzième siècles," *Bibl. Ecole Chartes* 26 (1875), 399.
18. Depoin, *Recueil St.-Martin-des-Champs*, 2:#285 ter.
19. Depoin, *Recueil St.-Martin-des-Champs*, 2:#341, #402, #429; 3:#515.

worth defending. The canons of Saint-Corneille of Compiègne successfully obtained from Louis VI in 1112 a charter confirming them in possession and enjoyment of their bridge. The document states that the bridge had belonged to them since the foundation of their monastery by Charles the Bald ca. 876. A royal provost had ruined their bridge, erected his own instead, and brazenly ignored his excommunication by the canons. Later the canons, with permission from King Philip I, rebuilt their bridge. Threatened with destruction by order of Louis VI, the bridge was saved when the canons were able to convince the king of the justice of their cause. He granted them the right to hold and possess the bridge without contradiction and with license and full power to build and improve it.[20] Surely the revenues from the bridge must have been considerable to warrant the exertions of the canons to retain ownership.

The same age that saw the concept of the bridge as a pious work score its most impressive triumphs witnessed also the proliferation of tolls on bridges. Even those spans in whose building motives of piety and public usefulness had been most to the fore became encumbered with tolls. At Tours in 1190 a document directed the equal division of the tolls by land and by water between the archbishop of Tours and Richard the Lion-Hearted in his capacity as count of Anjou.[21] At Albi at the end of the twelfth century the tolls on the bridge were going to the viscount, but after the Albigensian crusade the bishop inherited much of the viscount's power, including the right to receive the tolls on the bridge and the obligation to maintain it.[22] It seems probable that the greater interest in tolls was due partly to an increase in trade that rendered them more remunerative and partly to a greater demand for coinage, as the economy became increasingly a monetary one.

The donation of a bridge or of the tolls on it to a religious house was an acceptable present. Sometimes such a gift provides our earliest evidence for the existence of a bridge, as the present of half of the toll on salt on the bridge at Espalion to the monastery of Conques in 1060.[23] In 1016–1037 the monks of Saint-Germain-des-Près of Paris were given tolls on the passage across the bridge of Montereau and on boats passing the château, and at the end of the eleventh century Saint-Médard of Cappy received 60 *s.* of rent on

20. Morel, *Cartulaire de St.-Corneille de Compiègne*, 1:64.
21. L. de Grandmaison, ed., *Cartulaire de l'archevêché de Tours (Liber bonarum gentium)* (Tours, 1892–1894), 1:213–215, #LXXX.
22. A. Vidal, "Costumas del Pont de Tarn d'Albi," *Rev. des Langues Romanes* 44 (1901), 482.
23. G. Desjardins, ed., *Cartulaire de Conques en Rouergue* (Paris, 1879), p. 40, #572.

the crossing of the bridge at Warnéton (Nord).[24] Such gifts continued in the twelfth and thirteenth centuries. In 1169 the lepers of Popelin received 210 *sous* on the toll of the bridge of Pont-sur-Yonne,[25] about 1185 the monastery of Villarceaux obtained from Philip Augustus the right to collect a measure of salt from every boat passing under the bridge at Mantes on the Seine, and in 1209 the abbey of Fontaine-Daniel was given one-fourth of the passage over the bridge at Mayenne.[26] The reason for such a gift was ordinarily the benefit of the donor's soul. The document in which Fulk of Anjou, before 1116, gave his sluice at the bridge of Ponts-de-Cé to the convent of Fontevrault and to an individual stated: "The word of the Evangelist teaches us, saying, 'Lay up your treasures in Heaven'."[27] Between 1116 and 1125 the counts of Anjou gave the nuns of Fontevrault bridges at Chinon and at Ponts-de-Cé.[28] The gift of tolls on a bridge evidently sometimes involved no responsibility for upkeep, and the donation of a bridge was anticipated to provide a large income and negligible expenses for maintenance. Otherwise the gift would not have atoned for the sins of the donor.

Property rights played an important role in the history of the bridges of this period. The fear that construction of a new bridge would infringe on property rights frequently eclipsed completely the concept of it as a public improvement or a pious work. The Pont-de-la-Daurade was planned in the early twelfth century to be built by alms, but the count of Toulouse, who owned the watercourse, required an annual mass from the monks as recompense for his consent.[29]

Where the ownership of the watercourse was complicated by the existence of a ferry, the difficulties of bridgebuilders could be compounded. The reaction of ferry owners to the threat of a bridge was usually to demand com-

24. R. Poupardin, ed., *Recueil des chartes de l'abbaye de Saint-Germain-des-Près*, 1 (Paris, 1909), p. 86, #LIV. Depoin, *Recueil St.-Martin-des-Champs*, 1:#138. Warnéton (Nord), can. Quesnoy-sur-Deule, arr. Lille. Cappy, can. Bray, arr. Péronne (Somme).
25. M. Quantin, ed., *Cartulaire général de l'Yonne* (Auxerre, 1854–1860), 2:212, #CXCVI.
26. H.F. Delaborde, ed., *Recueil des actes de Philippe Auguste* (Paris, 1916), 1:193. Bienvenue, "Recherches," p. 562.
27. J. Chartrou, *L'Anjou de 1109 à 1151. Foulque de Jérusalem et Geoffroi Plantagenet* (Paris, 1928), p. 327, #6 bis.
28. Chartrou, *L'Anjou*, p. 353, #27; p. 357, #19. Also in 1116 the convent of Saint-Aubin was given by Hubert of Champagne a part of the tolls on the bridge at Durtal (Maine-et-Loire), built in the eleventh century: Bienvenue, "Recherches," p. 440.
29. Catel, *Histoire*, p. 196.

pensation. It was exceptional that the canons of Saint-Salvi of Albi were willing to give up their rights to tolls from their port in favor of free passage across the bridge for everyone "for the common improvement of the town and the utility of the whole Albigeois."[30] Such altruism was not shared by other ferry owners. Even at Avignon, the outstanding example of a bridge financed though alms, the proprietors of the port had to be bought out.

In some cases the difficulty of injured property rights was met by turning over to the ferry owner the new bridge. This was the case at Saumur in 1162 and at Chalonnes-sur-Loire in 1138–1148.[31] In the latter case the bishop of Angers received the bridge on the understanding that in case it were destroyed, the donor, the lord of Chalonnes-sur-Loire, was to receive the ferry tolls until such time as it should be rebuilt. At Saumur the transfer of the bridge tolls from the builders of the span, the knights and townsmen of Saumur, to the abbey of St.-Florent was made to compensate the abbot and monks for the loss of revenue from their ferry to the quarries of Saumur across the Loire. Evidently possession of customs of the bridge was extremely desirable. In 1264 the court of Charles of Anjou estimated that in a hundred years the abbey should have received 10,000 *livres* in tolls, that is, an average annual yield of 100 *livres.* In any case, the bridge tolls were so valuable that they exceeded the income from St.-Florent's ferry—the abbot and monks were obliged to promise to rebuild the bridge in stone, and the *Historia Sancti Florentii Salmurensis* claims that the abbot paid money in addition. The life of Frogerius Petitus de Sancto Loano, abbot of St.-Florent 1160–1174, states: "Pontem etiam de Salmuro, tradenti venerabili rege Anglorum Henrico, data tamen non modica quantitate pecuniae quae pro eodem ponte Turonis debebatur recepit."[32]

Sometimes the solution to the conflict of interest between bridgebuilders and ferry owners was the payment of an annual fee. Near Metz the abbot and monks of Saint-Arnoud owned a watercourse with a ferry at a point where the abbess of Bouxières-aux-Dames, owner of both banks, subsequently built

30. *Gallia Christiana*, 1, Instrumenta, no. 6, p. 4.

31. For Chalonnes-sur-Loire see Bienvenue, "Recherches," p. 212. For Saumur see E. Berger, ed., *Recueil des actes de Henri II*, Chartes hist. France, 7:365, and P. Marchegay, ed., *Archives d'Anjou* (Angers, 1834–1864), 2:172.

32. P. Marchegay and E. Mabille, eds., *Chronique des églises d'Anjou* (Paris, 1899), p. 311. Abbot Froger actively forwarded the material interests of the convent. He persuaded Pope Alexander III in 1164 that the provostship of Saint-Vivien-de-Pons should belong to Saint-Florent rather than to the bishop of Saintes. Abbot Froger also constructed a new cloister for the monks.

a bridge. The ensuing quarrel was settled in 1073 by a decision that every year on the feast of Saint Remi the convent of Bouxières-aux-Dames was to pay the abbot of Saint-Arnoud 12 *deniers*. If the bridge were destroyed by neglect or floods, the abbot was empowered to resume ferry service. If the abbess determined not to rebuild, she was to effect the removal of the piles so as not to interfere with the abbot's fishpond or his ferry.[33] In a similar case at Cahors in 1251 the bishop, owner of a port, demanded compensation for loss of revenue from his ferry because of the proposed bridge of the consuls. No friend to the idea of the bridge as a pious work, he was unimpressed by the consuls' explanation that the reason for constructing a second bridge over the Lot was "the utility of the town of Cahors and to avoid the danger of the boat and of crossing."[34] The bishop insisted he be allowed to maintain a collector on the new bridge to receive the toll from strangers, and that if the sum should amount to less than 10 *livres*, the commune must nevertheless guarantee him that figure annually.

There are other examples of protests against bridge construction, some of them on the simple basis that it should never interfere with existing revenues. This contention seems to have allowed no claims for a bridge either as a pious work or a public improvement. In 1181 there were complaints at Pont-sur-Yonne that the construction of a bridge had caused the loss of ferry tolls to the owners. These objections do not seem to have been taken seriously by the arbitrator of a quarrel over revenues at the town of Pont-sur-Yonne, perhaps because the bridge was a reconstruction of one that had been passable in 1169.[35] It was otherwise with the virulent protests of the bishop of Angers about 1136 against the new bridge and market of Geoffrey Plantagenet, count of Anjou, at Châteauneuf-sur-Sarthe (Maine-et-Loire). These innovations had adversely affected revenues from the episcopal market. To secure adequate compensation the bishop laid an interdict on Geoffrey's lands.[36] The action secured the bishop's ends, but the antithesis to the idea of the bridge as a charitable work has been reached.

Proprietary rights in bridges continued to exist throughout the remainder of the Middle Ages, but beginning in the middle of the twelfth century there were both an accelerating demand for more bridges and an increasing realiza-

33. A. Calmet, *Histoire ecclésiastique et civile de Lorraine*, 1 (Nancy, 1728), Preuves, p. 474. Bouxières-aux-Dames (Meurthe), can. Nancy-Est, Benedictine abbey founded in the tenth century.
34. Paris, B.N., Doat 118, fols. 116–117.
35. Quantin, *Cartulaire Yonne*, 2:332, #CCXIV.
36. Urseau, *Cartulaire noir d'Angers* pp. 286–287.

tion of the need for regular maintenance. A statement recognizing the fragility of bridges was included in a gift made by Henry the Liberal, count of Troyes, in 1157 to the collegiate church of Saint-Etienne of the free bridges situated on the domain of the church. The document contained the proviso: "If they are still standing a year and a day from now."[37]

At this time the idea came more and more to the fore that income from tolls involved responsibility for the bridge. For example, at Meaux the abbey of Saint-Denis was liable for the upkeep of the first arch of the bridge over the Marne, for which purpose tolls were collected on the bridge and market. The evidence comes from a document in which certain of the abbey's men purchased release in 1246 both from the tolls and from the obligation to maintain the arch.[38] The growth of the king's power and concomitantly of the royal bureaucracy had a hand in the more regular maintenance of bridges. In 1230 Saint Louis settled disagreements between two convents responsible for the upkeep of the Ponts-de-Cé. The monks of Saint-Aubin of Angers had possessed the parish of Ponts-de-Cé and four mills and fishponds under four arches of the bridge since the eleventh century. The nuns of Fontevrault had received the gift of all the rights of the counts of Anjou in the bridge at some time between 1115 and 1126, and in the thirteenth century they were responsible for repairs. The decision of Saint Louis provided that in case the four arches of Saint-Aubin fell, the monks were to have nine weeks to reconstruct them, after which they were to owe a fine of 20 *sous* daily to the abbess of Fontevrault. Destruction of the bridge by war or ice involved no fine for the monks, and if the entire bridge were destroyed, the monks were to finish their four arches after the nuns had completed the rest of the bridge.[39] The intervention of the royal government in determining responsibility for bridge maintenance was a portent of things to come.

A factor undermining the idea of a bridge as private property, that is, primarily a source of benefit or revenue to an individual or institution, was a change in bridgebuilders. In the eleventh and early twelfth centuries these had been almost entirely lay lords, bishops, and abbots with the collaboration at some places, as Albi and Toulouse, of the bourgeoisie. In the course of the twelfth century these dignitaries began to be replaced by townsmen. In northern France already in 1122 the men of Beauvais had bought from the bishop the bridges of that place and acquired from Louis VI the right to replace "the

37. T. Boutiot, *Histoire de la ville de Troyes et de la Champagne* (Troyes and Paris, 1870–1872), 1:212.
38. J. Doublet, *Histoire de l'abbaye de S. Denys* (Paris, 1625), 3:905–906.
39. Bienvenue, "Recherches," p. 215.

bridges and planks which they have on the water, whether they fall or are burned,"[40] and that without obtaining a license. At Metz in 1222–1223 at the same time that the bishop was constructing the Middle Bridge, the town was building the Pont-Thiffroy.[41] At Vermanton (Yonne) the townspeople built the bridge over the Yonne in 1238.[42] In southern and southeastern France late in the twelfth century, at Carcassonne, Agen, and Avignon, and elsewhere, burghers either built bridges or supported construction. At Lyons it was the citizens who initiated work on the Pont-de-la Guillotière ca. 1182 and who were in control of it before the brothers of the bridge. Also, in 1265 it was the people of the town of Saint-Saturnin-du-Port who went to the prior of Saint-Pierre to obtain his permission to begin construction of the Pont-Saint-Esprit.[43]

Bridgebuilding by the bourgeoisie was a move away from the concept of private property and towards that of public works, but it was not, in general, the same as the concept of bridges as pious works. The latter had advocated freedom of passage over the bridge for everyone, but the townspeople tended to insist on gratuitous crossing only for their own citizens. Thus in southern France at Romans the inhabitants had free passage across the Isère,[44] and at Millau in 1250 the residents were exempt from paying tolls on the Old Bridge over the Tarn.[45] Before the middle of the thirteenth century the acceptable alternative to tolls for bridges construction was not yet taxes (except for a few instances as at Carcassonne and Metz), but charity. The ideal (at least in southern France and along the Loire) was the *opus pontis*, the endowed bridge corporation supported by donations. Yet the demand for bridge construction and repairs was not being met by individual or corporate enterprise, whether for reasons of charity or of profit. Despite the antiquity of the concept of tolls as a perquisite, the demand began to be heard that the recipient of charges make the necessary repairs. As more and more bridges passed under the control of towns, the idea of the bridge as private property came under attack. To more and more people bridges seemed to fall in the province of government.

40. A. Loisel, *Mémoires des pays, villes, comté et comtes, evesché et evesques, commune et personnes de renom de Beauvais et Beauvaisis* (Paris, 1617), p. 266.
41. J. Schneider, *La ville de Metz aux XIIIe et XIVe siècles* (Nancy, 1950), pp. 13–14.
42. Quantin, "Histoire," p. 486.
43. P. Bruguier-Roure, ed., *Cartulaire de l'oeuvre des église, maison, pont et hôpitaux du Saint-Esprit (1265–1791), Mém. Acad. Nîmes*, ser. 7, vol. 12 (1889), Annexe, p. 3.
44. U. Chevalier, "Notice historique sur le pont de Romans," *Bull. Soc. dépt. Drôme* 2 (1867), 311.
45. J. Artières, ed., *Documents sur la ville de Millau* (Millau, 1930), p. 8, #18.

6

The Structure of Medieval Bridges to ca. 1250

THE MOST STRIKING FEATURE of the medieval bridge was probably its silhouette, studded with buildings and punctuated with pointed arches. The medieval bridge was distinctive both in its architecture and in its appearance, and a number of features were innovations representing an improvement over Roman practices. Not until the eighteenth and nineteenth centuries was there a sharp break with medieval bridgebuilding techniques. In fact, one of the best sources on the condition of medieval French bridges before nineteenth-century alterations is the seventeen folio volumes of Nodier, Taylor, and de Cailleux, *Voyages pittoresques et romantiques de l'ancienne France* (1820–1878).[1]

Undoubtedly the use of steel and the introduction of the suspension bridge have contributed to make medieval bridges appear old-fashioned. In the 1930s about the only masonry bridges being built in Europe were constructed by the Nazis, because Hitler's emphasis on guns caused steel for bridges to be in short supply. Masonry bridges built in the United States today commonly consist of only one or two arches.

It was quite otherwise in the medieval centuries and in the early modern period, up to and including the eighteenth century, when medieval bridges were admired both for their engineering and as esthetic achievements. The bridge over the Loire at Orleans was described in the fifteenth century as "one of the most beautiful jewels of the town."[2] Medieval bridge construction still commanded respect in the eighteenth century, and it is worth noting the opinion of Hubert Gautier, an architect and engineer. The three French

1. C. Nodier, J. Taylor, and A. de Cailleux, 17 vols. (Paris, 1820–1878).
2. Archives du Loiret, CC 967, fol. 3v.

spans he singled out for mention in his *Traité des Ponts* were all medieval and all crossed the Rhone. He wrote:

> Besides the Roman bridges we have modern ones that have their merits. One can count in France those of Avignon, of Saint-Esprit, and of Lyons over the Rhone. The first has been broken down: there remain only some arches on the side of Avignon. The second has survived entire; one can say that it is one of the most beautiful bridges in the universe.[3]

Finally, it should be pointed out that, although there are many medieval bridges still in use in France today, almost all of them are in the less frequented and poorer sections of southern and central France. Literary remains reveal that in the Middle Ages there were at least as many in other parts of the country, but that more prosperous sections could afford to replace bridges that were narrow, hump-backed, and never planned for fast traffic. They removed the old and built bridges more in accord with current notions of utility, beauty, and style.

Literary sources are the only evidence for the appearance of medieval French bridges before the year 1000, but it is clear that some of the fundamental features had already appeared before that date. One was the inclusion of fortifications in connection with bridges, as, for example, the towers built by 885 at the end of the Grand Pont and of the Petit Pont to defend Paris against the Northmen. Another was the location of mills below bridges (see Figs. 17, 18); there were mills at the Grand Pont in Carolingian times.[4] Typical of the whole medieval period was the construction of houses on bridges; they were mentioned in the sixth century as located from one end to the other on a bridge at Paris.

The fortified bridge was characteristic of the Middle Ages. Nevertheless, there were protests against this idea. In the first half of the eleventh century, at a time when castle building was condemned by assemblies of the Peace of God movement, the convents of Aniane and of St.-Guillem-le-Désert provided that there should never be a tower or church raised on their bridge at the Gouffre-Noir over the Hérault.[5] However, the idea of the bridge as a pious work was not enough to inhibit the construction of fortifications, which continued to be built on bridges and at bridgeheads throughout the Middle Ages.

3. H. Gautier, *Traité des ponts*, 4th ed. (Paris, 1765), 1:5.
4. *Gallia Christiana*, 7, Instrumenta, no. 18, cols. 16–17.
5. C. Devic, J. Vaissète, *Histoire générale de Languedoc*, new ed. (Toulouse, 1872–1893), 5:393, #CLXV. Similarly, an act of the chancery of Frederick I forbade fortifications on the bridge at Bonpas without consent of the bishop of Avignon: Archives départementales de Vaucluse, G5, #38.

At Lyons open warfare between the archbishop and chapter of Lyons resulted in the construction of two bridge towers before 1208, and at Rouen in 1204 there was a barbican at the head of the bridge.[6] The military importance of bridges was bound to result in their destruction. In 1153 during a quarrel between Henry II and the citizens of Limoges, the king broke down the bridge,[7] and in 1205 the bridge at Angers was set afire by John Lackland fleeing from Philip Augustus.[8]

There is much evidence for the location of houses on bridges, some of it derived from accounts of floods. In an inundation of 1175 "many houses on the bridge at Angers fell because of the violence of the waters," according to the Annals of St.-Serge.[9] When ice broke down the bridges at Saumur and Tours in 1235, a chronicler, with the exaggeration characteristic of some of his kind, tells us: "Apud Turonem pro ruptione pontium submersi fuerunt homines infiniti."[10] Houses on bridges were a profitable source of income. To help finance bridge maintenance at La Rochelle before 1202 Master Isembert, the builder, had constructed dwellings *circa pontem*, and the tenants were to pay an annual *cens* of 6 *sous* for the repair of the bridge and its illumination. In 1207 the owners of the bridge, the brothers of the Aims House and the brothers of the Temple of La Rochelle, were planning the joint construction of new houses.[11]

Throughout the Middle Ages a bridge was considered a desirable site for houses, shops, and mills. At Paris the existence of the Grand Pont is attested by the gift of a mill in 1070,[12] and in the next two centuries there is ample

6. V. Mortet and P. Deschamps, *Recueil de textes relatifs à l'histoire de l'architecture, XIIe–XIIIe siècles* (Paris, 1929), p. 210. When the bourgeoisie of Rouen made their submission to the king of France in 1204, they agreed to turn over to Philip Augustus the barbican at the head of the bridge so he could construct a fortress there. They offered to destroy four arches of the bridge toward Rouen, whenever it should please the king, and either to construct a gate there or to wall it up, as the king might prefer: Rigord, "De gestis Philippi Augusti Francorum regis," *Historiens de France*, 17:58.
7. "Fragmentum genealogicum ducum Normanniae," *Historiens de France*, 18:242.
8. J.M. Bienvenue, "Recherches sur les péages angevins aux XIe et XIIe siècles," *Le Moyen-Age* 63 (1957), 214.
9. Louis Halphen ed., *Recueil d'annales Angevines et Vendomoises* (Paris, 1903), p. 104.
10. Annales de Saint-Florent, in Halphen, *Recueil*, pp. 125–126.
11. Mortet and Deschamps, *Recueil, XIIe–XIIIe siècles*, pp. 194–195. L.M. De Richemond, "Chartes de la commanderie magistrale du Temple de la Rochelle (1139–1268)," *Arch. Saintogne* 1 (1875), 36.
12. M. Prou, ed., *Recueil des actes de Philippe I, roi de France (1059–1108)* (Paris, 1908), p. 143, #LIII. The authencity of a document in which King Henry I of

evidence for houses, mills, and shops on the Grand Pont and the Petit Pont,[13] the only two bridges at Paris at this time, as had apparently been the case since antiquity. Louis VII decreed in 1141 that on the Grand Pont and on it alone the moneychangers should carry on their affairs.[14] Already about 1138–1141 houses on the Petit Pont were occupied by scholars of Paris. About this time John of Salisbury became acquainted with Adam of Petit-Pont, so named from the location of his school.[15] Guy of Bazoches about 1175 mentioned the two bridges in a glowing account of the city "raised above others by the royal diadem."[16] To the north the Grand Pont was described as wealthy, crowded with people eager to buy, and as abounding in boats and merchandise: "Behold, this place has no equal. However, the Little Bridge is dedicated to passers-by, strollers, or disputants in logic."[17] Both bridges at this time had a connection with scholars, for in 1198 Giles of Paris in enumerating in a poem the men of letters, including professors and scholars, born in Paris and vicinity, mentioned Adam of the Grand Pont and Pierre of the Petit Pont.[18] Godfrey, canon of Saint-Victor, wrote in a poem that on the Petit Pont a venerable order of old men, eminent for their knowledge and morals, taught the multitude.

The roadbeds of both bridges were paved, and Godfrey's poem states that the Petit Pont, built of cubic stones, stood as solidly as though the piers had been bronze pillars and would never fall (this before the disaster of 1206, when the Petit Pont's three arches were carried away and many of its houses overturned by a flood).[19] An anonymous canon of Sainte-Geneviève in his *Ex miraculis sanctae Genovefae* describes the inundation of 1206 which swept away the three arches of the Petit Pont, mentioning that the stones had been

France ca. 1033 confirmed the possessions of Saint-Magloire, including a mill at the Grand Pont, has been questioned: R. Lasteyrie du Saillant, ed., *Cartulaire général de Paris* (Paris, 1887), p. 115, #87. See J. Guérout, "Le Palais de la Cité à Paris, des origines à 1417," *Mém. Féd. Paris* 1 (1949), 124, n. 3.

13. Lasteyrie du Saillant, *Cartulaire*, p. 275, #286; p. 276, #287; p. 470, #578.
14. Lasteyrie du Saillant, *Cartulaire*, p. 277, #288.
15. R.L. Poole, "The Masters of the Schools at Paris and Chartres," *Studies in Chronology and History* (Oxford, 1934), pp. 224, 235. Later Adam became bishop of Ansaph: H. Rashdall, *The Universities of Europe in the Middle Ages*, ed. F.M. Powicke and A.B. Emden (Oxford, 1936), 1:288.
16. C.H. Haskins, *The Renaissance of the Twelfth Century* (Cambridge, 1927). The quotation is from the Meridian paperback edition (Cleveland, 1955), p. 380.
17. E. Berger, "Description de Paris vers 1175 par Guy de Bazoches," *Bull. Paris* 4 (1877), 39.
18. *Histoire littéraire de la France*, 9 (Paris, 1750), 75, 77, 78.
19. *Historiens de France*, 18:798. See also Rigord, in *Historiens de France*, 17:61.

laid up with mortar.[20] This was typical of medieval bridges, and it is thought that the Petit Pont had been rebuilt by the bishop of Paris, Maurice de Sully, ca. 1175.[21] Godfrey mentions parapets on the Petit Pont to protect the passers-by, who could view the river from holes in the wall.

On many a medieval bridge the houses were perched on the piers and on them alone, and if the houses extended beyond the piers, they were propped up with timbers braced against the sides of the pier. At some places, as at the Ponte Vechio in Florence, the buildings extended from one end of the bridge to the other and were supported by it. However, in medieval Paris houses extended back from the bridges and were supported on piles driven into the stream bed. Upstream from the Grand Pont the river belonged to the abbey of Saint Magloire[22] and upstream from the Petit Pont to the bishop of Paris, in each case as far as the Ile Saint-Marie (the present-day Ile Saint-Louis). In 1212 the bishop sold to a widow Oudarde "six *toises* (11.694 m.) of our water on the upper side of the Petit Pont" to build a dwelling "behind the house of the said Oudarde."[23] She sold 10 *toises* (19.49 m.) to four men planning to erect dwellings, but it is uncertain just how much frontage each of these men was to have.[24] It is understandable that the dimensions of the houses along the roadbed should be less than their depth upstream or down. At the Grand Pont, too, there was crowding, for in 1249 the dimensions of one workshop were 2.7 meters wide by 3.8 meters deep, and of another 8.77 meters wide by 11.69 meters deep.[25]

Surviving French medieval bridges testify to the originality of their builders. Medieval contributions (as distinct from Roman) included the introduction of novel shapes for arches and cutwaters, an improvised method of founding piers, and the use of a sand-lime mortar and of small stones (except where large stones were readily available from ancient Roman structures).

20. "Ex miraculis sanctae Genovefae," *Historiens de France*, 18:797–798.
21. "Ex chronologica Robert monachi Sancti Mariani Altissiodorensis," *Historiens de France*, 12:298.
22. J. Tardif, *Monuments historiques [Cartons de rois], 528–1789* (Paris, 1866), p. 225, #404. In 1120 Louis VI gave to the abbey of Saint-Magloire the exclusive right to fish in the Seine between the Ile-Saint-Marie (Ile-Saint-Louis) and the Grand Pont, with the exception of the mill race of the bishop of Paris. Later this privilege included the ownership of the river between these points: Lasteyrie du Saillant, *Cartulaire*, p. 215, #194.
23. B.E.C. Guérard, ed., *Cartulaire de l'église de Notre-Dame de Paris* (Paris, 1850), 4,1: 142.
24. H. Sauval, *Histoire et recherches des antiquités de la ville de Paris*, 1 (Paris, 1724), 216.
25. Guérard, *Cartulaire de Notre-Dame*, 4,2:470–471.

Another novelty lay in the adoption of a more substantial distance between the keystone and roadbed. Moreover, at many bridges the roadbed, rather than being level, sloped upward from each river bank toward the center. Medieval bridges showed a variety in the shape of the cutwaters that Roman bridges had not. In the Middle Ages the upstream cutwaters were pointed, and the downstream cutwaters might be pointed, square, or even entirely missing. Cutwaters might be low, or they might rise to the height of the springing of the arch or to the top of the parapets.

Another area of medieval innovation lay in the shape of the arches. The Roman arches in France had been semicircular; those of the medieval period were semicircular, segmental,[26] or pointed. Each of these arches has its advantages and disadvantages. The semicircular arch is relatively stable and easy to build; but in floodtime the higher the water level, the less the space available for navigation. The segmental arch exerts more pressure against the piers, but when properly balanced it allows the use of small piers and involves little blocking of the freeway. However, the chief advantages of the segmental arch were difficult to realize in the Middle Ages, for in an age which could not calculate thrusts and strains, large piers were almost inevitable. The pointed or ogival arch is very stable, and in floodtime there is less question of decreasing the area of the freeway as the waters rise; but the use of pointed arches calls for a greater number of piers in the river than does the semicircular or segmental arch.

It has been assumed that the ogival arch was introduced into bridge construction from the example of church builders. Although the pointed arch had been employed in Syria several hundred years earlier, its first appearance in what is now France seems to have been at Cluny ca. 1095.[27] If the pointed arches in medieval bridges were copied from ecclesiastical structures, this would mean that all ogival arches in bridges should date after about 1100, and this appears to have been the case. The ogival or two-centered arch produced greater stability, but at no time in the Middle Ages was its superiority sufficiently clear-cut so that it superseded its competitors, and round arches continued to be built throughout the period. Moreover, the pointed arch generally disappeared from bridge construction in the course of the sixteenth century.

26. In architecture a segmental arch is an arch of which the intrados forms the segment of a circle, meeting the imposts at an angle.
27. N. Pevsner, *An Outline of European Architecture* (Harmondsworth, England, 1943), p. 75.

In France the extant medieval bridges are located in the southern and central sections of the country, where circumstances have favored their survival. For the builders of these bridges the Roman spans of southern France were of great importance as models. The Romans used massive piers and employed the semicircular arch, but they varied the width of the arch as convenient. This required that the springing of the arch should be higher or lower according to its width, so as to retain the round arch and the level roadbed. The stones were hewn stones, and the masonry was laid up without mortar, although the stones sometimes were fastened together by iron bars sealed in lead. The arches were composed of parallel courses of stones without interlocking, apparently because during construction this method made possible the re-use for each successive course of the same centering. Cutwaters were square or lacking downstream, and upstream they were generally pointed and rose in some cases to the height of the springing of the arches.

Examples of Roman structures available for imitation by medieval builders included the ancient bridge at Sommières (Fig. 2) and the aqueduct at Pont-du-Gard (see Fig. 6). In the Roman bridge at Sommières each pier was perforated by an opening designed to carry off flood waters and to lessen the weight of the structure. The width of the piers (3 m.) was more than a third of that of the arch openings (8.5 m.). The bridge at Sommières underwent repairs in the Middle Ages, and the medieval masonry is laid up with mortar having narrower courses of stone than those used by the Romans. In 1844–1845 the bridge was widened, and sidewalks were put in, but even now the width is only 6.5 meters between the parapets.[28]

The Pont-du-Gard, probably dating from the first century A.D., was originally built as an aqueduct to carry water across the Gardon on its way to Nîmes, and it was used as a footbridge during the Middle Ages (Fig. 6).[29] At the Pont-du-Gard the upstream cutwaters are pointed, and their top is higher than the springing of the arches.[30]

The next bridges still to be seen in France date from the eleventh century, a gap of some 800 years. The absence of such monuments from the Merovingian and Carolingian period is not accidental; it accurately reflects a period of disorder when few bridges were built. The ones that were built were in danger

28. W. Emerson and G. Gromort, *Old Bridges of France* (New York, 1925), p. 31.
29. *Dic. top. Gard.*, s.v.
30. Emerson and Gromort, *Old Bridges*, pp. 25–28. C.S. Whitney, *Bridges* (New York, 1929), pp. 69–70. For the Pont-Ambroix (Hérault), a ruined Roman bridge, see A. Grenier, *Manuel d'archéologie gallo-romaine*, 2 (Paris, 1934), 190.

of man-made destruction, even if they escaped damage from floods and the weather. Some evidence on the character of the early Grand Pont has been provided by a discovery made in 1855 during excavations for a sewer. Inside the abutment of the Pont-au-Change on the right bank there came to light a pier with two adjoining arches, identified as vestiges of the bridge of Charles the Bald in Théodore Vacquer's sketch and article in the *Revue archéologique.*[31] There was a strong Roman air to the remains of the bridge. The arches were semicircular, and the upstream cutwater, pointed with a flat top, extended above the springing of the arches, so that the whole bore a definite resemblance to the Pont-du-Gard. The roadbed was laid almost on the stones of the extrados, a Roman feature, and the pier was excessively thick (2.75 m.) in relation to the arch opening (5.35 m.), a ratio of more than one to two. The interior of the pier was composed of rubble in the medieval manner, and the roadbed was low, as was common in bridges of the Middle Ages—only 8.23 meters above the record-breaking low water levels of 1719 and well beneath later flood levels.

The date of the pier and arches uncovered at the end of the Grand Pont has been in dispute. Vacquer declared that the remains are proof that the site of Charles the Bald's bridge was the same as that of the Grand Pont in the thirteenth century. On the other hand, he remarked that the pier had a marked batter and displayed a sophistication in the laying of the masonry which he would not have expected in a ninth-century bridge. The refinement of technique encouraged Jean Guérout[32] to reject a Carolingian date and refer the remains to the twelfth century, when, he presumed, building practices were more advanced. There is a difficulty with Guérout's argument. We lack evidence to determine the typical level of Carolingian competence in building stone bridge piers. Since a careful examination of Vacquer's description and sketch shows a number of distinctively Roman features and few medieval ones, it is probable that the pier and arches were built earlier, rather than later. They must have been constructed after the middle of the fourth century, when Emperor Julian described the bridges of Paris as built of timber, and they could well be from the ninth century, when we know that at least some of the main bridge during the Norman siege was of masonry. Since tradition and documentary evidence confirm the site of the Grand Pont as the

31. T. Vacquer, "Lettre sur la découverte d'une partie du Grand Pont," *Rev. arch.* 12 (1855), 502–507.
32. Guérout, "Le Palais," p. 138. See also C. Enlart, *Manuel d'archéologie française depuis les temps mérovingiens jusqu'à la Renaissance*, 2 (Paris, 1903), 267.

Fig. 1. Caesar building a bridge across the Rhine. British Museum, Harl. 6205, fol. 21, from the sixteenth century.

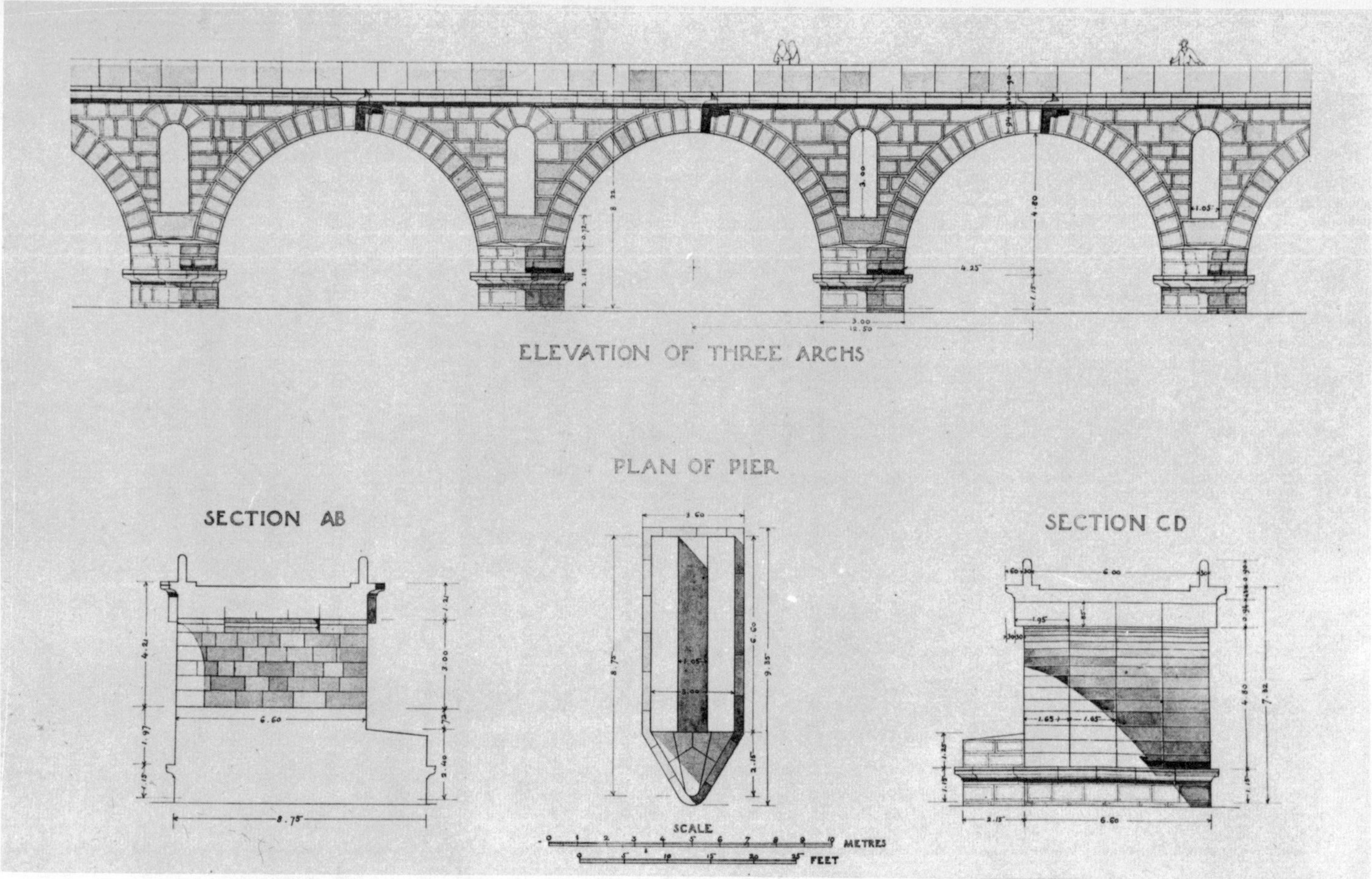

Fig. 2. Roman bridge at Sommières. From W. Emerson and G. Gromort, *Old Bridges of France* (New York, 1925), P1. V.

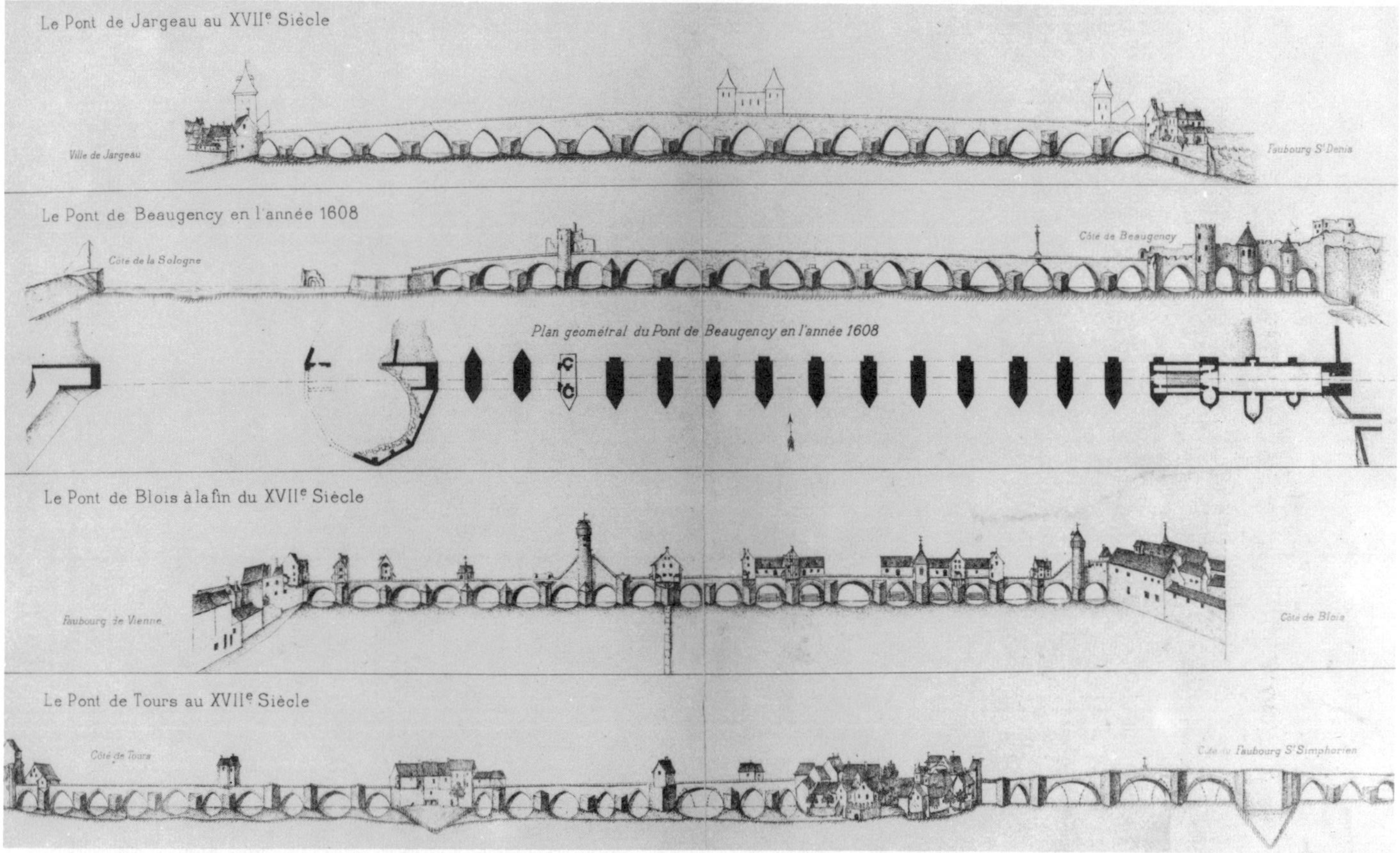

Fig. 3. Seventeenth-century drawings of the bridges at Jargeau, Beaugency, Blois, and Tours. From A. Collin, *Le Pont des Tourelles à Orléans* (Orleans, 1895), Atlas, P1. VI.

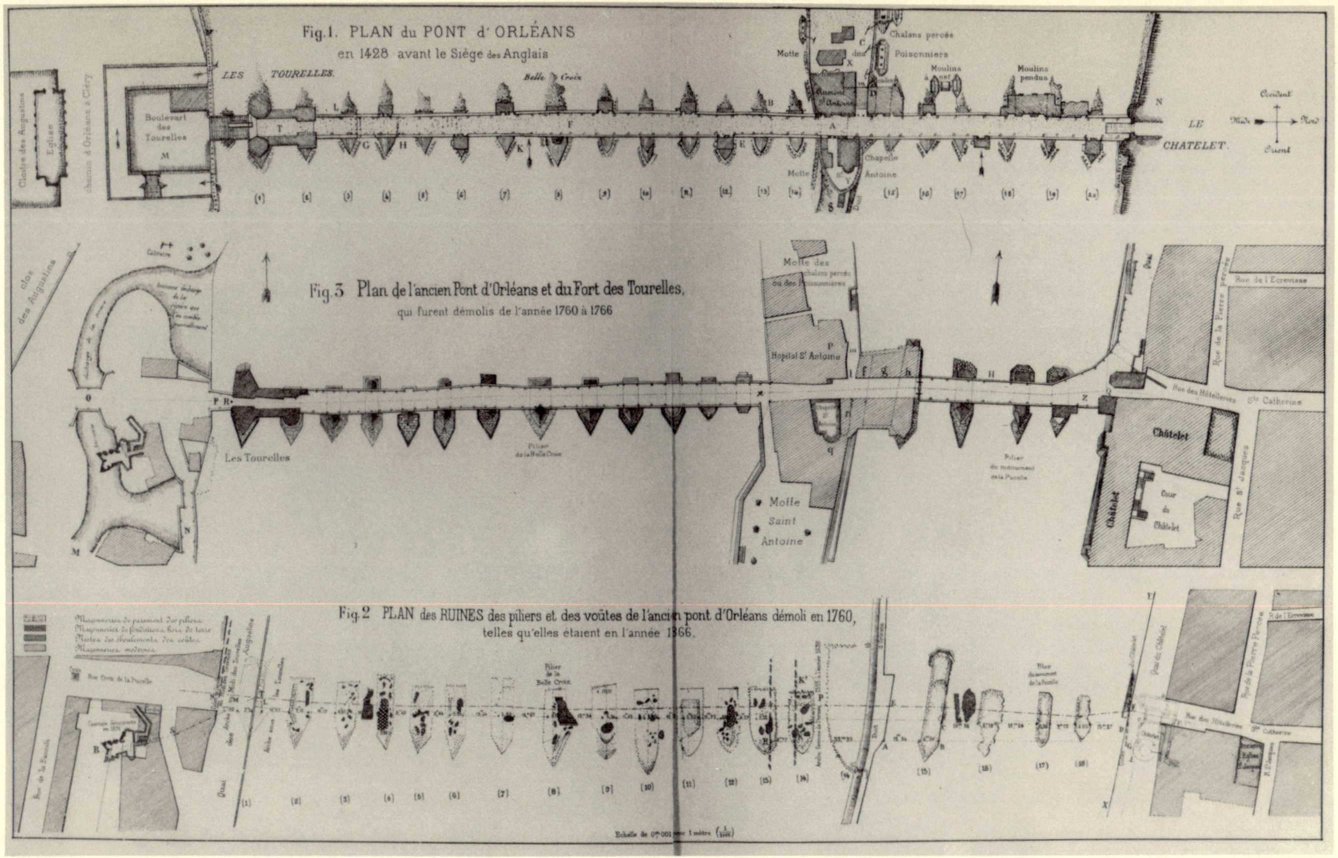

Fig. 4. Plans of the medieval bridge at Orleans. From Collin, *Le Pont des Tourelles*, Atlas, Pl. II.

Fig. 5. Espalion. From C. Nodier, J. Taylor, and A. Cailleux, *Voyages pittoresques et romantiques dans l'ancienne France* (Paris, 1820–1878), vol. 2, Pl. 77.

Fig. 6. Pont-du-Gard. Photo supplied by French Embassy, Press and Information Division.

Fig. 7. Pont-Saint-Bénézet, Avignon. Note the resemblance in the manner of laying up the stones of the voussoir to that of the Pont-du-Gard in Figure 6. Drawing by Vytautas Sakalauskas.

Fig. 8. View from the roadbed of the bridge at Avignon showing the chapel of Saint-Bénézet. Drawing by Jesse Kapili.

Fig. 9. The Perussis Altarpiece. View of the bridge at Avignon by an unknown French painter ca. 1450–1490. Note that one of the arches has been replaced by wooden scaffolding and that the artist has omitted the chapel. New York, Metropolitan Museum of Art, bequest of Mary Wetmore Shively in memory of her husband, Henry L. Shively.

Fig. 10. Eighteenth-century view of the Pont-Saint-Esprit. Caisse Nationale des monuments historiques.

Fig. 11. Montauban. Caisse Nationale des monuments historiques.

Fig. 12. View of Albi from the cathedral tower, showing the bishop's palace in the foreground and looking upstream toward the medieval bridge in the center of the picture. From a negative of Lindsley F. Hall, courtesy of the School of Architecture, University of Oregon.

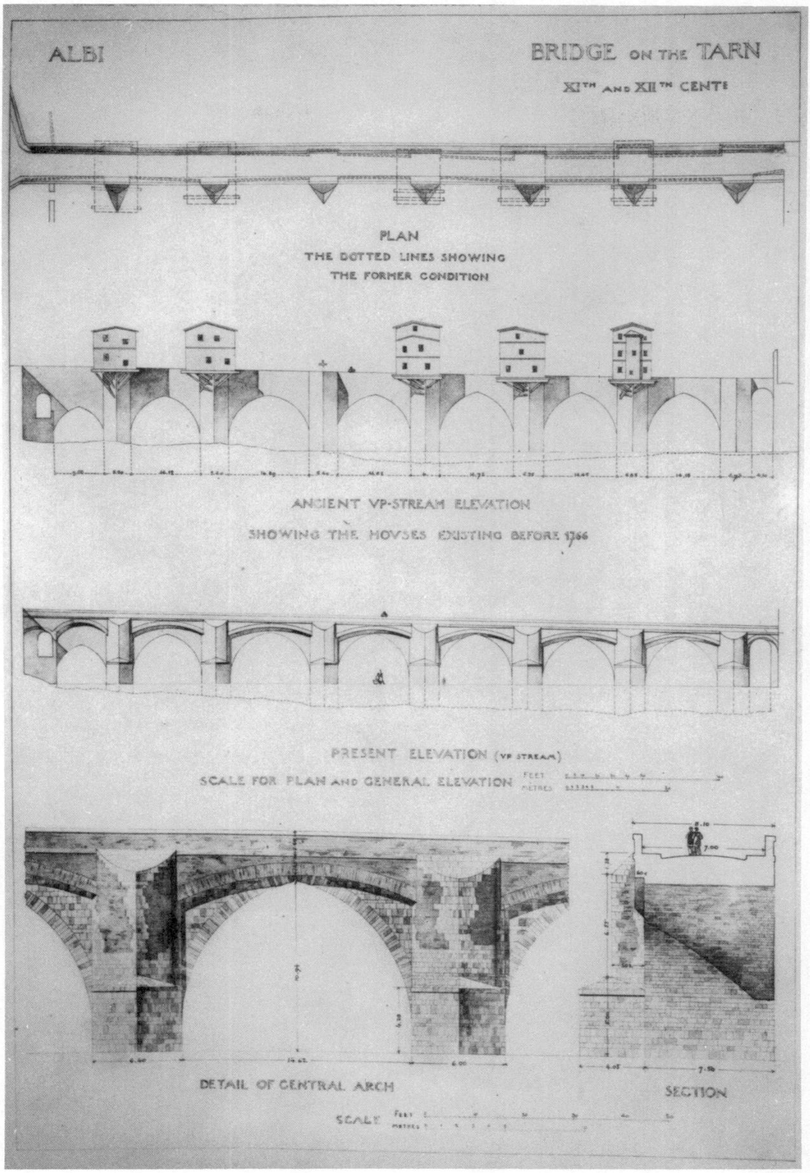

Fig. 13. Plan of the bridge over the Tarn at Albi. Emersón and Gromort, *Old Bridges of France*, Pl. VI.

Fig. 14. Cahors. From Nodier, Taylor, and Cailleux, *Voyages pittoresques*, vol. 1, part 2, Pl. 71 bis.

Fig. 15. Seals of the town of Lyons depicting a bridge, from 1271 and 1320. From M.–C. Guigue, *Cartulaire municipal de la ville de Lyon* (Lyons, 1876), on flyleaf.

Fig. 16. A bridge at Paris ca. 1317. Paris, B.N., MS fr. 2091, fol. 99r.

Fig. 17. Mills downstream from a bridge at Paris, ca. 1317. Paris, B.N., MS fr. 2092, fol. 37v.

Fig. 18. Mills at Corbeil in the time of Louis XI. Paris, Archives Nationales, S2116.

site of the bridge of Charles the Bald, it seems reasonable to attribute a certain sophistication to ninth-century bridgebuilders.

To know what one end of a bridge was like in the medieval period tells us little about the rest of it. Medieval bridges frequently consisted in part of stone arches and in part of a timber roadbed supported at some points by masonry piers and at others by piles. According to Abbo's epic, there were at least some stone piers, but parts of the bridge, probably the roadbed, must have been of timber to account for the Northmen's attempt to burn the structure.[33] Certainly some arches in the middle of the river must have had wider openings, because this was customary in Roman and in medieval bridges, as for example at the Pont-du-Gard, so as to allow for the passage of the stream and of boats. Besides, an arch opening of only 5.35 meters would barely have furnished passage for Viking boats the width of the Gokstad ship of the year 900—that is, 5.3 meters.[34] One at least of the arches of Charles the Bald's bridge must have been wider to account for the Northmen's desire to pass under the bridges of Paris.

Literary evidence gives us a description of a tenth-century bridge in a state of neglect after the breakdown of provisions for methodical maintenance. Richer tells us in his *History of the Franks* that while traveling from Reims to study at Chartres in 991 he reached the bridge at Meaux just at dusk. Its ruinous condition was such that citizens of the town hesitated to cross in broad daylight. In vain Richer searched the banks of the stream and inquired of the residents to procure a boat. In the end he was obliged to attempt a crossing of the bridge. Fortunately he was accompanied by an experienced traveler who covered the gaping holes in the roadbed with his shield or with planks to enable the horses to pass safely. When one such gap had been negotiated, his companion moved the planks on to the next, and in this manner they were able to cross the bridge without incident.[35]

Romanesque bridges, as the name implies, strongly resembled Roman structures, and one of the earliest medieval examples is still to be seen on the Gouffre-Noir on the Hérault. Sometime between 1031 and 1048 an agreement to build a bridge at that point was concluded between the convents of Aniane and St.-Guilhem-le-Désert.[36] An eighteenth-century drawing, made before the bridge was widened and repaired, shows four semicircular arches,

33. Abbo, *Le Siège de Paris*, ed. and trans. H. Waquet (Paris, 1942), book 1, line 378.
34. J. Brøndsted, *The Vikings*, trans. K. Skov (Baltimore, 1965), pp. 140–141.
35. Richer, *Historiae Francorum* 4.50, ed. and trans. R. Latouche (Paris, 1937), p. 225 ff.
36. Devic and Vaissète, *Histoire*, 5:393, #CLIV.

one much larger than the others, and no cutwaters downstream.[37] These were Roman features, but above the voussoir there was a second ring of stones concentric with it, atypical of Roman style. Moreover, the document states that the builders were to provide, among other materials for construction, lime, sand, iron, and lead. To fasten stones together with iron bars sealed in lead was standard Roman practice, employed at the Pont-du-Gard, but the use of sand-lime mortar was, as we have said, a medieval innovation.

A Roman feature that was rarely copied during the Middle Ages was that of piercing the upper part of the pier with an opening, but bridges at Béziers and at Montauban (Fig. 11) nevertheless have this characteristic. At Béziers the structure is probably of the late twelfth century, although first attested in 1209.[38] The arches were semicircular, the main arch opening being larger than the others and situated under the highest point in the roadbed. The fact that the roadbed rose to a point and sloped down again (which was true also at Millau, built before 1156)[39] was a typically medieval feature, but at Béziers the cutwaters, pointed upstream and square down, rise to a point just below the opening through the pier, an ancient practice.

Cutwaters pointed both upstream and downstream seem to have been a medieval innovation. It has been suggested in the twentieth century that the cutwater pointed both upstream and downstream was a technological improvement,[40] protecting the pier from eddies below the bridge. In the early sixteenth century a mason employed by the town of Albi to inspect the bridge deplored the fact that all the piers lacked cutwaters on the downstream side, so that the piers suffered unnecessarily from the current.[41] The presence of pointed cutwaters did not necessarily eliminate this hazard, for in 1859 M. Aymard,[42] engineer at the Pont-Saint-Esprit, reported that boats managing to pass safely under the arch on the left bank were caught by an eddy caused by the enormous piers. Whatever the merits of the pointed downstream cutwaters, they failed to convince most builders of medieval French bridges, who continued to use square downstream cutwaters.

37. D. Rey, *Etudes archéologiques sur le vieux Millau*, 2 (Millau, 1923), 12.
38. E.E. Viollet-le-Duc, *Dictionnaire raisonné de l'architecture française*, 10 vols. (Paris, 1854–1868), 8:229 ff.
39. Rey, *Etudes*, 2:20–21. The length of the bridge was 218 m., the width of the first eight arches varied from 11 m. to 15 m., and the breadth of the pier was 4.5 m.
40. E.B. Mock, *The Architecture of Bridges* (New York, 1949), p. 16.
41. E. Jolibois, "Le Vieux Pont d'Albi," *Rev. Tarn* 2 (1878–1879), 200.
42. M. Aymard, "Notice sur les travaux executés au pont Saint-Esprit pour la construction d'une arche marinière," *Annales des Ponts et Chaussées*, 3rd ser. (1859), p. 2.

Medieval features of the bridge at Carcassonne included cutwaters that were pointed upstream and down, rising to the height of the parapets, as well as the great thickness of the bridge between the voussoirs and the roadbed. A new and distinctive characteristic of the bridge at Carcassonne is that it has segmental arches instead of the Roman semicircular arches. Admittedly the divergence is small, as they are only 1/15 lower than they would have been had they been semicircular.[43] Nevertheless, the idea of segmental arches was employed both at the Pont-Saint-Bénézet and the Pont-Saint-Esprit, the extant arches of the latter dating certainly to the second half of the thirteenth century.

Any discussion of French medieval bridges centering on extant structures is necessarily a discussion of stone bridges. Yet it is probable that in the eleventh and twelfth centuries most bridges were constructed of timber. For example, in 1120 a bridge over the Aveyron, presented to the abbey of Moissac, was described as consisting of planks on piles.[44] It has been doubted whether the expression "lapideus pons," appearing frequently in the chronicles, meant anything more than that there were stone piers. Specifically, a charter of Fulk Nerra, count of Anjou, mentioned that in 1028 he built a "lapideus pons" at Angers.[45] Yet, some 120 years later when the span was washed out, the Angevin Chronicle described it as "ex maxime parte ligneum."[46] It is probable that both statements are correct. Again and again in the medieval period it happened that one or more stone arches of a bridge collapsed, whether from faulty construction, neglect or floods. The original structure was then patched with timber.

The difficulty of maintaining bridges is well illustrated at Paris in the twelfth and thirteenth centuries. Floods carried away the bridges in 1196/7, 1206, 1280, and 1296. Part of the roadbed of the Grand Pont was of timber construction even in 1280 before the flood, and in 1284 the cathedral chapter of Paris replaced some wooden planking on the Grand Pont with a stone arch.[47] The bridge was reconstructed in timber after the inundation of 1296,

43. E. Degrand, *Construction* (Paris, 1888), 2:60. Eleven "arches semblables de 11 m à 14 m d'ouverture en arc de cercle, surbaissées de 1/15 à peine." J. Needham, *Science and Civilisation in China*, vol. 4, part 2 (Cambridge, 1965), p. 226, in giving the seventh century as the date of the earliest Chinese segmental arch bridges, states that 1340 was the year of the oldest European bridges of this type.
44. A. Champollion-Figeac, *Droits et usages concernant les travaux de construction* (Paris, 1860), p. 127. Paris, B.N., Doat 131, fol. 241r-v.
45. Bienvenue, "Recherches," p. 213.
46. Bienvenue, "Recherches," p. 214.
47. Guérard, *Cartulaire Notre-Dame*, 4,2:491–494, #CXXIV.

and the former bridge, despite the fact that it had been partly of stone, partly of wood, was referred to in the fourteenth century as the stone bridge.[48] The ruins of the piers, stretching from the Palace to the Châtelet, remained visible for centuries, until finally in the early seventeenth century the Grand Pont or (as it was then called) the Pont-au-Change, was reconstructed in stone. The vestiges of one of the old piers was described by Sauval in the eighteenth century. There remained six courses of calcareous limestone fastened together with iron bars to form a pier 4.87 meters long by 2.59 meters wide. The pier, which had a facing of dovetailed stones and a filling of rubble, was pointed in one direction and square on the opposite side. Sauval concluded that in olden times builders did not drive piles or put platforms under piers and that this was a recent invention.[49]

A number of great twelfth-century bridges had piers founded on starlings rather than on piles. London Bridge and bridges at Blois and Orleans were constructed with this type of pier (Figs. 3, 4). The method of construction consisted in driving short piles into the riverbed to a depth of perhaps half their length, so as to form an enclosure. Into this pen rubble was dumped to a height above water level, beams were laid down, and the masonry of the pier was then built on this starling, the top of which was paved. Investigations of the remnants of the old bridge at Orleans, torn down in 1760, have confirmed this statement. Alexandre Collin examined in 1866 the vestiges of the piers in the river bed with a view to ascertaining the method of founding piers. He confined his investigations to that part of the river between the Tourelles and the site of the Motte Saint-Antoine, for this was the oldest part of the bridge, almost certainly dating from the twelfth century. The stretch between Saint-Antoine and the Châtelet was largely rebuilt after February 1434/5. Collin found that in several cases at the site of piers there were still a few short piles, some upright, some horizontal, and rubble to the height of a meter. Above this there was masonry extending, at low water, to within .80–1.5 meters of the surface.[50] Collin's conclusion that the piers of the old bridge at Orleans had been placed on starlings was confirmed in September 1943. Then very low water permitted close examination of the second pier of the old bridge counting from the north bank, one that very probably dated

48. Gustave Fagniez, *Etudes sur l'industrie et la classe industrielle à Paris au XIIIe et au XIVe siècle* (Paris, 1877), p. 168, n. 2.

49. H. Sauval, *Histoire et recherches des antiquités de la ville de Paris* (Paris, 1724), 1:225.

50. A. Collin, *Le pont de Tourelles à Orléans (1120–1760)* (Orleans, 1895), pp. 401–403.

from the twelfth century. There were remains of an enclosure of piles, a series of horizontal sleepers under masonry, with large stones around the periphery and small ones inside the pier.[51] It is not known when medieval builders invented the starling, but it was certainly in use in the twelfth century, and perhaps the eleventh, as Blois seems to date from the eleventh. In the late thirteenth century there is firm evidence for piers founded on driven piles rather than on starlings.

There are a number of bridges first attested in the Romanesque period, the silhouettes of which seem to indicate that extensive changes were made at later dates. Among these are the bridges at Albi (Figs. 12, 13), Espalion (Fig. 5), Blois, and Tours (Fig. 3). Plans to erect the bridge at Albi were made about 1035, and the bridge must have been constructed at some time prior to 1157, when the suburb across the river was called Bout-du-Pont.[52] The first evidence that the bridge was in use comes from 1178.[53] Today it has one semicircular and seven ogival arches of brick and stone. The last arch was built in modern times, apparently to replace the drawbridge. The shape of the arches at Albi would seem to preclude a date for their construction in masonry earlier than about 1100. Beginning in the fourteenth century, the history of the bridge is well-documented and shows repeated repairs. For example, in the years from 1403 to 1410 three of the eight arches were reconstructed.[54]

The bridge at Espalion (Aveyron) also presents a contradiction in style between the way it presumably looked when first attested in 1060 and its present-day appearance (Fig. 5). Now two of its four arches of red sandstone are somewhat pointed and a third very slightly so, while the fourth arch, built in 1724, is round.[55] The bridge has a slope of ten percent to a point over the main arch, but the three towers the bridge once carried have been removed, and the structure has obviously been widened. It is possible to date the bridge at Espalion more accurately by the shape of the cutwaters than of the arches. The cutwaters rise abruptly to the height of the parapets and are pointed

51. G. Chenesseau, "Note sur une des piles de l'ancien pont d'Orléans," *Bull. Orléanais* 24 (1943), 473 ff.
52. "Glanures historique," *Rev. Tarn* 8 (1890–1891), 212.
53. Jolibois, "Le Vieux Pont d'Albi," p. 197.
54. M.N. Boyer, "Rebuilding the Bridge at Albi, 1408–1410," *Technology and Culture* 7 (1966), 27.
55. G. Desjardins, ed., *Cartulaire de Conques en Rouergue* (Paris, 1879), p. 402. F. de Dartein, *Etudes sur les ponts en pierre remarquables par leur décoration antérieures au XIXe siècle*, 1 (Paris, 1912), 35 ff. Whitney, *Bridges*, p. 97.

both upstream and down, a combination found in a number of bridges built between the middle of the twelfth and the middle of the thirteenth centuries: Millau (before 1156), Carcassonne (before 1184), Orthez (before 1254), and Entraygues-sur-le-Lot (1269).[56] It seems very likely that the bridge at Espalion assumed more or less its present form between the years 1150 and 1250.

The bridge at Tours (Fig. 3), like that at Espalion, apparently was first built in the eleventh century. It was planned by Count Eudes of Blois in 1031–1035, attested in 1108, and torn down in the eighteenth century.[57] Despite its early date, seventeenth- and eighteenth-century pictures show it with a majority of its arches ogival, a shape of arch probably not used earlier than the beginning of the twelfth century. The bridge was damaged by floods in 1235 and 1309, and in 1313 it was called "the great stone bridge over the Loire."[58] The bridge continued to be altered over the years. Due to the menaces of the Hundred Years' War in 1367–1368 the bridge wall was broken to install a drawbridge.[59] There are differences in outline and the number of piers in pictures of the seventeenth and eighteenth centuries, by which time it had become fashionable to remove a pair of ogival arches and substitute one large round arch to facilitate navigation.[60] Probably the bridge was rebuilt in stone in the middle of the twelfth century, when so many stone bridges were being built. The fact that the method of founding piers is similar to that used at Orleans, built before 1176, is an additional argument for assuming a comparable date.

The old bridge at Blois, mentioned in 1078, seems to have survived into the modern period with most of its stone arches intact. A seventeenth-century picture (Fig. 3) shows all the arches as round with the exception of two ogival ones, presumably added later. The changes over the years were chiefly in the cutwaters. The pointed upstream cutwaters vary: some are low with a pyramidal capping, some rise to a point above the springing of the arch and are flat on top, as at the Pont-du-Gard, and some ascend vertically to the height of the parapets.[61]

56. Dartein, *Etudes*, 1:29, 43, 44.
57. Indre-et-Loire, *Inventaire-sommaire des archives départementales antérieures à 1790*, vol. 8, ser. H (Tours, 1891), p. 73.
58. E. Boutaric, ed., *Actes du Parlement de Paris* (Paris, 1863–1867), 2:123, #4275. L. de Grandmaison, ed., *Cartulaire de l'archevêché de Tours* (Tours, 1892–1894), 1:304–305.
59. J.M.A. Delaville-Le-Roulx, ed., *Registres de comptes municipaux de la ville de Tours* (Tours, 1878–1881), 1:136, 366; 2:220, 224, 270–271, 277–279.
60. Collin, *Le Pont*, pp. 203–204.
61. Collin, *Le Pont*, Atlas, Planche VI.

Dissimilar arches in a given medieval bridge are now common. Examples are the Pont de Saint-Nicolas-de-Campagnac (Gard), serviceable by 1261, which has nine arches, some pointed and some circular, and the bridge at Saint-Généroux (Deux-Sèvres), traditionally of the thirteenth century, which has a pointed arch at each end and three semicircular arches in the middle.[62] Two bridges at Limoges,[63] the Pont-Saint-Martial and the Pont-Saint-Etienne, have ogival arches, but there have been alterations over the centuries. The Pont-Saint-Martial has lost its wooden tower, and the parapet has been built of smaller stones than the lower parts of the bridge. Evidently the Pont-Saint-Martial (like the Pont-Saint-Bénézet) was originally built without a wall to protect traversers from falling off the bridge. The cutwaters, pointed upstream and square down, climb abruptly to the tops of the parapets. Consequently, they offer the advantage of providing a roadbed wide enough to permit one vehicle to turn out for another.

Vicissitudes of a bridge are illustrated typically by the history of the bridge at Beaugency. Much repaired, it is still in use. The chartulary of Notre-Dame-de-Beaugency contains a document, dated between 1160 and 1182, which mentions the bridge as receiving part of the toll on the Loire. In 1203 the lord of Beaugency presented to the abbey the fifth arch of the bridge, so that it might build a mill and fish pond there. A drawing of the bridge at Beaugency made in 1608 shows the southern end as missing due to a flood; but it also indicates structures which in the twentieth century are no longer there (Fig. 3). There were two stone towers, presumably with a fortified gate, on the seventeenth pier from the northern side. On the second pier was a chapel dedicated to St. Jacques, and on the third pier a tower.[64] In the seventeenth century almost all the arches were pointed. The low cutwaters were pointed upstream and square down with the exception of two which were pointed in both directions.

In accounting for the stylistic changes in a medieval bridge, it should be remembered that the people of the period rarely had at their disposal at any one time adequate funds for bridge construction. Hence they ordinarily replaced only those parts requiring immediate attention. Sometimes they modernized the shape of an arch, sometimes they imitated the old style, but

62. Enlart, *Manuel*, 2:296. Emerson and Gromort, *Old Bridges*, p. 49.
63. F. de Verneilh, "Architecture civile du moyen-âge. Construction des ponts," *Annales arch.* 20 (1860), 100, 107.
64. G. Vignat, ed., *Cartulaire de l'abbaye de Notre-Dame de Baugency*, 16 (Orleans, 1879), 124, 146. Collin, *Le Pont*, p. 218; Atlas, Planche VI.

they rarely felt any compulsion to achieve symmetry. By the seventeenth century a great bridge, such as those at Orleans or Tours, represented a succession of styles in the shape of its arches and cutwaters.

Most of the great medieval bridges over the Loire were built between 1000 and 1250, and they lasted into the eighteenth century, when new levees along the banks raised the flood level to a height of 7 meters above mean water level. This was fatal to bridges built to accommodate a crest of 5–5.5 meters, and one after another the giants at Blois, Jargeau, and Tours succumbed (Fig. 2). Although the bridge at Orleans, due to incessant care, maintained a tenuous existence, it had to be replaced in 1750–1760.[65] There remains only about half the medieval bridge at Beaugency. As engineering achievements to meet a certain set of conditions, these bridges had been eminently successful. They lasted with repairs for 600 to 700 years–until altered circumstances rendered them obsolete.

65. R. Dion, *Le Val de Loire, étude de géographie régionale* (Tours, 1934), p. 396.

III

FROM CA. 1250 TO CA. 1500

BRIDGES AS PUBLIC WORKS

7

Financing Bridges

Governmental Assistance (ca. 1250–ca. 1350)

IN THE COURSE of the thirteenth century, as in Carolingian times, it became widely accepted throughout what is now France that bridges were public works, that is, that they were built primarily for the general benefit, that their support should consist of tolls and taxes, and that there construction and maintenance were a governmental responsibility. In the first half of the thirteenth century once again it is in the South that we find the verbalization of the concept. An arbitral award of 1239 involving the Pont-de-la-Daurade of Toulouse stated that bridge construction and repairs were matters of "public utility and common convenience."[1] Hence the tendency for more and more bridges to come under the control of towns and fewer and fewer to continue to be the property of the clergy or the nobility, held for reasons of defense, income, or charity. The initiative of towns was assisted by the royal government of France and in the Agenais by that of England. Not only did the king of France aid towns by granting extraordinary tolls and taxes, but the courts also pinpointed responsibility for maintenance of particular bridges.

Three famous bridges built in this period and still to be seen are the Pont-Saint-Esprit, the bridge at Montauban, and the Pont Valentré at Cahors. Of these only the first was financed by donations. After the middle of the thirteenth century, despite the shining example of the Pont-Saint-Esprit, most bridges were built with funds raised from taxes and tolls. This state of affairs does not seem to have curtailed significantly the conviction that bridges were a work pleasing to God and that gifts for their construction conferred spiritual benefits. In the fourteenth century the inhabitants of Vienne and collectors at Pont-Saint-Esprit solicited alms for bridges over the Rhone. Legacies and requests for indulgences continued during the late thirteenth and fourteenth centuries. Thus in 1308 Pope Clement V granted for seven

1. Archives de la Haute-Garonne, H Daurade 8 (1239).

years an indulgence of 100 days to those faithful who, truly penitent and confessed, stretched forth a helping hand to the fabric of the bridge the Dominicans were building near Nîmes.[2] Indulgences were provided for four wooden bridges at Grenade-sur-Garonne in 1309, for a bridge at Sully-sur-Loire in 1318, and about the same time for the Pont-Saint-Esprit.[3] The receipt of tolls did not mean that those in charge of bridges were willing to neglect the recourse to charity. At Agen authorizations had been received again and again to collect tolls and taxes for the bridge, but the financial burden was great, and in 1346 the consuls dispatched a delegation to the pope to request indulgences.[4]

At this time gifts and legacies continued to be given to bridges for the same reason they were made to churches. It was for the love of God that the commune of Prats-de-Mollo (Pyrenées-Orientales) in 1321 made a donation of 15 *livres* of Barcelona to the *opus pontis* over the Tech near Céret.[5] The bridge at Marcigny (Saône-et-Loire) in 1315, 1318, and 1320 and the *opus pontis* at Vieille-Brioude (Haute-Allier) in 1340 received legacies.[6]

By the middle of the thirteenth century it was already clear that although charity was obviously the ideal method of financing bridges, it was not about to provide them in the requisite numbers. The government moved on several fronts to remedy this lack. One was to insist that a region tax itself to provide funds, and here the case of Agen is an illuminating example. Originally, in 1189, at the same time as work was going forward on the Pont-de-la-Guillotière at Lyons and the Pont-Saint-Bénézet at Avignon, the citizens of Agen had attempted to finance a bridge across the Garonne through donations. The effort failed, and local enthusiasm waned. Only the intervention of the English government assured resumption of work. In 1283–1285 the vice-seneschal of the Agenais convoked a regional assembly to arrange financing

2. *Regestum Clementis Papae V ex Vaticanis archetypis* (Rome, 1884–1888), #2689.
3. *Regestum Clementis Papae V*, #4344. G. Mollat, ed., *Jean XXII (1316–1334). Lettres communes* (Paris, 1904), 1: #6643. L. Bruguier-Roure, *Cartulaire Pont-Saint-Esprit, Mém. Acad. Nîmes*, Ser. 7, vol. 17 (1894), Annexe, pp. lviii–lxix, 175, 178, 182–185. For indulgences granted to donors to the bridge over the Lot at Entraygues (Aveyron) in 1269, see Paris, B.N., Doat 172, fol. 311 ff.
4. *Inventaire-sommaire arch. com. Agen*, BB 16, p. 9.
5. Testamentary gifts were made in 1326 and 1334: P. Séjourné, *Grandes Voûtes*, 6 vols. (Bourges, 1913–1916), 1:10.
6. Marcigny (Saône-et-Loire), arr. Charolles. L. Blin, "Le Grand Chemin de Paris à Lyon par la vallée de la Loire au bas moyen âge (de Décize à Marcigny par la rive gauche)," *Comité des Travaux hist.* (1958), p. 252 ff. For Vieille-Brioude see Séjourné, *Grandes Voûtes*, 2:15.

of the bridge. Having secured the support of the inhabitants, in 1286 King Edward I of England authorized the collection of a *barragium* or *barra* of one *denier bordelais* on each cavalier and one obole on each pedestrian to build the bridge in stone.[7] (The advantage of a *barra* or *barragium* was that it could be collected at any point, whether or not the bridge was passable.) In addition to the toll, a tax was instituted. King Edward in 1289 ordered his seneschal to convoke a general court of the whole Agenais, consisting of clergy, barons, knights, consuls of all the towns, and other inhabitants of the region, to approve a hearth tax of six *deniers arnaudins* to complete the bridge at Agen. They were to do this at the instance of the king and the divine will, and for their own honor, lest the money already disbursed should have been spent in vain and "lest so praiseworthy a work should remain unfinished to the dishonor of the whole countryside."[8]

After the institution of the hearth tax the initiative in building the bridge passed to the town of Agen. The consuls (probably about 1298) hired Bertrand Tichender, a native of Cahors and resident of Moissac, to build the bridge for the sum of 30,000 *sous arnaudins.* This amount proving insufficient, a lawsuit procured the contractor an additional 200 *livres.* A stone bridge was built, but it was washed out before 1320. The consuls repeatedly appealed to the royal government, sometimes in England, sometimes in France, for authorization to continue to levy tolls and taxes. In 1299 the consuls sent to Paris, in 1320 to King Edward II, and in 1324 to Charles IV of France for renewal of the *barra.* A new tax was authorized by Philip VI in 1329, to be collected only with the inhabitants' consent. The *souquet*, a charge on wine sold and on grain ground in the town and suburbs of Agen, was to go to the rebuilding of the bridge for fifteen years. The income was applied sometimes to the bridge, sometimes to fortifications, sometimes to other purposes.[9]

There are extant a list of *barra* charges from about 1298 and an inquiry made by a judge and royal councillor in 1335 into the receipts of the *barra* and the expenditures for the construction of the bridge. Cavaliers traversing the bridge paid 1 *denier arnaudin*, and pedestrians one *maille*, half as much. More than half the income of the *barra* came from charges on boats belonging

7. *Inventaire-sommaire arch. com. Agen*, DD, p. 4. When the people of the neighboring town of Condom protested the *barra*, the consuls of Agen agreed to their permanent exemption on the payment of the sum of 100 *livres arnaudins.*
8. C. Bémont, ed., *Rôles Gascons* (Paris, 1906), 2:442.
9. C. Bémont, *Rôles Gascons*, 2:443, 444. *Inventaire-sommaire arch. com. Agen*, CC, p. 5.

to non-residents passing under the bridge. For the years 1331–1335 the total income from the *barra* was 1271 *l.* 2 *s.* 1 *d.*, of which 1234 *l.* 2 *s.* 11 *d.* had been spent on work on the bridge.[10]

Undoubtedly the willingness of the French royal government, after the middle of the thirteenth century, to grant the right to collect tolls and taxes for the purpose of bridgebuilding was largely responsible for the very great number of medieval bridges built in the second half of the century. Philippe de Beaumanoir in his *Coutumes de Beauvaisis* observed that no one was permitted to institute tolls except the king, who was allowed to establish a "false custom" for a time for the expense of repairs to bridges, roads, and churches.[11] For example, the royal government ordered the collection of a temporary toll for the repairs of the bridge at Melun. When in 1273 the convents of Saint-Port and of Barbeaux protested that they were exempted by royal charter, the Parlement of Paris disallowed the objections on the grounds that the false custom, imposed for a certain time, was for the common good, as well theirs as that of all passing through.[12]

Government grants of permission to levy a *barra* for bridge construction and maintenance were frequent. The Parlement of Paris in 1275 allowed the consuls of Cahors to levy a toll for the repair of the great stone bridge for two years,[13] and in the fourteenth century the Pont Valentré at Cahors was partially financed by a *barrage.* The first stone of the latter was laid on 17 June 1308. Upon request Philip IV aided the commune in the improve-

10. C. Higounet, "Un Compte de ferme de la barre d'Agen au début du XIVe siècle," *Annales du Midi* 62 (1950), 351–355. The number of barriers varied. In 1331–1332 there were four, one for boats plus "las barras del Molinar e del Temple e de la Granero," but in 1332–1333 there were five, for "la barra de la Reclusa" had been added.
11. P. de Beaumanoir, *Coutumes de Beauvaisis*, ed. A. Salmon (Paris, 1900), 2:262–263.
12. E. Boutaric, ed., *Actes du Parlement de Paris* (Paris, 1863), 1:176, 328. *Les Olim*, ed. A.A. Beugnot (Paris, 1839–1848), 1:929. Barbeaux (Seine-et-Marne), arr. Melun, cant. Châtelet, com. Fontaine-Le-Port. At Melun the royal treasury received substantial suns from "the passage of the arch of the bridge," namely, 208 *livres parisis* in March 1299, 152 in May, and 444 in June: J. Viard, ed., *Les Journaux du trésor de Philippe IV le Bel* (Paris, 1940), cols. 355, 398, 399, 437. The provost of the merchants and alderman of the city of Paris in 1348 brought suit before the Parlement of Paris against the mayor, peers, and sworn men of the commune of Pontoise to protest the collection for three years of a toll on merchandise passing under the bridge for its repair. The merchants lost their case, and the custom to rebuild the bridge was collected: G. Huisman, *La Jurisdiction de la municipalité parisienne de saint Louis à Charles VII* (Paris, 1912), p. 90.
13. Boutaric, *Actes du Parlement*, 1:190, #2075; 333, #222.

ment of the road from the town to the Pont Valentré, situated outside the walls, where it formed an advanced defense, commanding the course of the Lot and the plain next to the town. In 1323 Charles IV granted the commune a *barra* to be applied exclusively to the construction of this bridge, and to procure its renewal the consuls dispatched one of their number to Paris in May 1328. The bridge was completed by 1355.[14]

At Riom (Puy-de-Dôme) royal grants of permission to levy a *barra* for bridge repairs and other needs were made in 1303, 1318, and 1325.[15] At Tours because of the flood of 1309 there was a new tax levied: "l'emolument du naviage de la roupture du grant pont de pierre de Loyre de Tours." By an agreement of 1313 between the royal provost and the archbishop, half the receipts were to go to repair the bridge and the other half to the church of St. Martin of Tours.[16] At Toulouse in 1313 "the latest inundations and poverty" were given as the reasons why even though the town was supposed to pay the expenses of repairing bridges, it was unable to do so, and the king gave the commune permission to levy a *barra* for two years.[17] The king of France granted the town of Millau permission to levy a *barra* to reconstruct the New Bridge, already old enough in 1286 to need repairs. In addition, the toll on the Old Bridge was to go for this purpose, and two years later the clergy agreed to pay *tailles* for bridge construction.[18] At Villeneuve-sur-Lot (Lot-et-Garonne) the inhabitants secured from Edward I in 1289 the renewal of their *barra* for four years, "considering the intolerable expenses which the consuls and *universitas* of Villeneuve, diocese of Agen, have had for the construction of a bridge over the Lot," and because of public utility.[19] The bridge, begun in 1282, is still to be seen, although the main arch was rebuilt in the seventeenth century.

The government also aided bridge maintenance by fixing the responsi-

14. Pierre Martin was gone for forty-six days. His chief expenditures were for the hire of his horse, food for himself and his valet, and the seal validating the royal letters: F. de Dartein, *Etudes sur les ponts en pierre remarquables par leur décoration antérieures au XIXe siècle*, 1 (Paris, 1912), 12–14.
15. Philip VI in 1331 ordered the consuls of Riom to apply the *barra* to rebuilding the Pont de Nonette, which had formerly brought in 60 *livres* annually and over which no one had passed in the last ten years: *Inventaire-sommaire arch. com. Riom*, CC 58, p. 74.
16. Boutaric, *Actes du Parlement*, 2:123, #4275.
17. *Inventaire arch. com. Toulouse*, AA 5, 1:85.
18. J. Artières, ed., *Documents sur la ville de Millau* (Millau, 1930), p. 27. See also the statement for 1288, pp. 30, 53.
19. Bémont, *Rôles Gascons*, 2: #1588.

bility for individual structures. Parlement insisted that the heirs of the original builder should repair the bridge. The Parlement of Paris decided in 1291 that since in 1254 Helias, lord of Bergerac, had left 36,000 *sous* to build a bridge for the benefit of his soul, responsibility for it devolved on his heirs, not on the commune.[20] The same principle was followed in Roquecourbe (Tarn) in 1317, where the consuls were successfully able to prove that since the bridge had been built by the countess's ancestor, she was obliged to maintain it ("stones as well as wood"), although they made a gratuitous donation of 200 *livres* to rebuild it.[21] In somewhat the same situation Charles, count of Anjou, in 1264 brought into court the monks of Saumur for their failure to honor their promise to Henry II to reconstruct the bridge in stone. The judgment condemned them to pay 500 *livres* annually until the bridge was finished, but it is not known whether the sentence was actually carried out.[22] Here should be noted a change in the climate of opinion. No longer it is enough for a convent to pray for the soul of the donor; if the monks have been given a bridge, they are obligated to maintain and improve it.

At some places the principle was beginning to be established that the recipient of tolls was to be held responsible for repairs to the bridge. Thus the Parlement of Paris decided in 1260 that Lord Guillaume de Villerets should repair the bridge at Villerets (Eure), because the custom of the country required the owner of the tolls to maintain the bridge.[23]

Legal enforcement of the obligations of ownership was unwelcome to some proprietors of bridges. For example, the convent of Fontevrault, which early in the twelfth century had thankfully received the gift of the bridges of Ponts-de-Cé from the counts of Anjou, now willingly exchanged them for other property. The abbess and nuns turned over to Charles of Valois in 1293 everything they owned between one bridgehead and the other—the tolls and justice, "together with the houses and the islands and other things" they held.[24] The abbess explained that the maintenance of the structure was too expensive; every year it had been necessary to journey six leagues to obtain wood for its repair. All the abbess retained of her rights in the bridge was that of free passage for herself, the nuns, and their servants.

20. Boutaric, *Actes du Parlement*, 1:437, #775. F. de Verneilh, "Architecture civile du moyen-âge," *Annales arch.* 16 (1856), 296.
21. "Glanures historiques," *Rev. Tarn* 7 (1888–1889), 240.
22. J.M. Bienvenue, "Recherches sur les péages angevins aux XIVe et XIIe siècles," *Le Moyen-Age* 63 (1957), 217.
23. *Les Olim*, 1:496, Villerets (Eure), com. Ecouis.
24. P. Marchegay, *Archives d'Anjou* (Angers, 1843–1854), 2:265.

The disillusionment with bridges as a source of profit is well documented in the late thirteenth and fourteenth centuries. It was exceptional that the prior of La Sône (Isère) was willing to purchase the chapel, hospital, and bridge at that place in 1323.[25] Far more typical were the experiences of the bishop of Albi and of the prior of Le Ris (Nièvre). The bishop of Albi, who had acquired the bridge there along with other powers of the viscount after the Albigensian Crusade, found that receipt of the tolls failed to compensate him for the maintenance of the bridge, and in 1269 he turned it over to the town, retaining only the right to receive the tolls on the occasion of an annual saint's day.[26] The prior of Le Ris received in 1343 the handsome sum of 40 *livres* as income from his bridge across the Allier, but ten years later the bridge had disappeared. In 1353 the prior was involved in a lawsuit over the ferry from which the convent had formerly derived 10 *l.* annually. It was to avoid claims on the part of two lords to the port that the prior of Le Ris originally had built the bridge shortly before 1343.[27]

The disappointment of convents with the administration of bridges is exceptionally fully told in records at Lyons. When the administration of the *opus pontis Rhodani* was taken away from the brothers of the bridge by the archbishop of Lyons about 1308, it was transferred from laymen to the abbey of Hautecombe of the order of Cîteaux in the diocese of Lyons to be rebuilt in stone. The fabric of the bridge consisted of two chapels, the hospital, almonry, houses, and other property, and the bridge itself. Brother Etienne repaired the wooden bridge, but on becoming abbot of Hautecombe he asked to be relieved of an impossible task—that of reconstructing in stone the Pont-de-la-Guillotière—and by an act of 1314 the *opus pontis* was conveyed to the abbey of Chassagne. Evidently the abbot, with an eye to the bridge's endowment, expected it to be a profitable venture, for he paid the convent of Hautecombe 1000 *livres viennois* to reimburse it for its expenses.

The abbot, however, had not adequately reckoned with the River Rhone and the citizens of Lyons, for the stream continually undermined and carried away parts of the structure and the townsmen demanded a functional bridge. The citizens obtained from Philip V in 1320 permission to levy for five years

25. J.P. de Valbonnais, *Histoire de Dauphiné* (Geneva, 1722), 1:286; Preuves, pp. 197–198.
26. A. Vidal, "Costumas del Pont de Tarn d'Albi," *Rev. des Langues Romanes* 44 (1901), 482.
27. A. Bruel, *Visites des monastères de l'ordre de Cluny de la province d'Auvergne aux XIIIe et XIVe siècles* (Paris, 1891), pp. 8–9, 37, 44. On the subject of the disillusionment with bridges, see also de Valbonnais, *Histoire*, 1:266.

a *barra* to be collected on the bridge, in Mâcon, and at places near Lyons. All sums were to be received by two or three citizens chosen by the seneschal of Lyons as *probi viri* and to be applied in their entirety to the rebuilding of the bridge.[28] The citizens complained to the pope of the administration of the abbot of Chassagne, who, they claimed, was allowing the extensive property of the bridge, donated by the faithful, to be alienated to his friends. Pope John XXII in 1328 transferred care of the bridge to a committee and appointed three ecclesiastics to investigate charges that the abbot of Chassagne was failing to apply the income of the bridge and alms of the faithful to the purposes for which they had been intended. The result, claimed the citizens, was that the Pont-de-la-Guillotière was hazardous to cross and in imminent danger of collapse.[29] A royal document of 1332 granted the collection of a *barra* at Mâcon, stating as a reason "la ruyne et faiblesse du pont de Lyon sur le Rosne."[30]

A settlement of the problem was made in 1334 by the archbishop of Lyons with the consent of the chapter, of the abbot and convent of Chassagne, of the abbot of the monastery's mother house, and of the consuls of Lyons, since, as the document explained, what affects all should be approved by all. The deed asserted that the income from rents and charity was inadequate to finance even repairs to the bridge, let alone reconstruction in stone, and that the *opus pontis* was a heavy burden on the monastery. Nevertheless, the abbey of Chassagne had built half a stone arch. Since no one is obliged to perform the impossible, inasmuch as the archbishop cannot consent to the ruin of a convent in his own diocese, and since such a burden can better be supported by the public, the archbishop decided to still popular clamor and the excitement of the citizens of Lyons. The convent having requested to be relieved of the burden of the bridge, the archbishop determined to separate the spiritual and the temporal. The consuls of Lyons were to have the bridge, the chapel erected on it, and the old house of the almonry and dependent buildings, but the other chapel and the hospital with the curtilage were to go to the abbey of Chassagne. To support the service of the chapel and hospital the archbishop transferred to them from the *opus pontis* five properties with houses in the city, and one farm and two vineyards (one with a house) outside Lyons.[31] The bridge thus found itself denuded of most of the property

28. M.–C. Guigue, "Notre-Dame de Lyon," *Mém. Lyon* (1874–1875), pp. 224–226.
29. M.–C. Guigue, *Cartulaire municipal de la ville de Lyon* (Lyons, 1876), p. 22 ff.
30. Guigue, "Notre-Dame de Lyon," p. 227.
31. Guigue, "Notre-Dame de Lyon," p. 228; *Cartulaire municipal*, p. 169 ff.

acquired over a period of 150 years through the generosity of the faithful, and this was done in favor of two objects formerly considered very minor parts of the *opus pontis*—a small overnight shelter (the hospital) and one of the bridge chapels.[32] The archbishop's reasoning was that a hospital and chapel were properly objects of charity but that the bridge was a matter of public concern, the responsibility of the commune of Lyons, to be supported by taxes and tolls.

The history of the Pont-de-la-Guillotière in the fourteenth century illustrates a number of trends in the development of bridges. The most important tendency was for the control of bridges to pass to towns. At Avignon the bridge had gone from the possession of the brothers of the bridge directly to the commune; at Lyons it went from the brothers of the bridge to one convent after another and then to the town. Just as so many abbeys had received bridges as gifts only to find that their maintenance was an intolerable burden, so the convents of the Lyonnais gladly relinquished control of the Pont-de-la-Guillotière. Another trend was the decline in the concept of the bridge as properly an object of charity. This tendency may have facilitated the diversion of parts of bridge income or endowment to ecclesiastics, as happened at the Pont-Saint-Bénézet and the Pont-Saint-Esprit, as well as at the Pont-de-la-Guillotière.

The conclusion that the bridge was a public responsibility to be financed by taxes and tolls was an idea which had been popular since the middle of the thirteenth century (although it had been advanced earlier), and the appeal to the government for permission to levy a toll or a *barra* was by this time the usual procedure. For example, in 1265 the queen of England made the *bastide* of Montségur (Gironde) responsible for the upkeep of bridges, and in 1282 the lord of La Roquebrou (Cantal) recognized the right of the community of the bourg to impose on those among them who consented to it a *taille* for the construction and maintenance of the bridge, of the church, of the *enceinte*, and of paved roads.[33] In 1281 King Edward I granted to the men

32. To maintain the service of God in the chapel there were to be two monks and a servant, and to care for the poor in the hospital two servants. A minute listing of the requirements in food, wine, and supplies necessary to the chapel and hospital shows the latter to have been primarily a small overnight shelter, where the annual sustenance provided for the poor was less than that for the two monks and three servants. The explanation of why the abbey was to retain one chapel was that it had formerly been in charge of this alone.

33. J.F. Samazeuilh, *Histoire de l'Agenais, du Condomois et du Bazadais* (Auch, 1846–1847), 1:332, Montségur (Gironde).

of Condom for three years the right to collect a *barra* to repair the town bridge.[34]

If there was difficulty in financing bridges in the last half of the thirteenth and first half of the fourteenth centuries, part of the reason was the insistence of local residents on exemption and the application of tolls to purposes other than bridge repair. For example, at the end of the thirteenth century when the timber bridge at Pont-sur-Yonne was in poor repair, the Parlement of Paris decided that the chapter was obligated to rebuild it, although the king received the tolls.[35] At Millau in this period the tolls of the New and Old Bridges were only applied to maintenance in times of emergency. In 1317 an annual rent of 130 *livres* from the toll on the Pont Vieux went to the hospital of Saint John of Jerusalem, and the remainder to the viscount of Fézenzaguet.[36] However, on 12 March 1338 (n. st.) the king gave to the town for two years the toll on the Pont Vieux to finance repairs on the New Bridge.[37] The next year, according to an agreement, the king was declared to be owner of the bridge and a toll collector was stationed on it. The tolls now went to the hospital of Saint John of Jerusalem, the viscount of Fézenzaguet, and to the king, but the townspeople of Millau were responsible for repairing the bridge, fortifying, and guarding it. When crossing the bridge, the inhabitants were exempt from all charges on goods or supplies for their own use, and those who paid taxes, or took part in the watch day or night, or guarded the gates, were completely free of all charges.[38] Bridge repairs were financed by taxes to which all inhabitants of Millau were forced to contribute.[39] We lack evidence that at this time the townspeople were dissatisfied with this arrangement.

The determination of local residents to achieve free passage cut income from bridge tolls. For example, the abbess of Sainte-Glossinde of Metz insisted on gratuitous crossing over the bridge at Auboiré, a nearby village

34. The charges were to be one *denier* on a loaded beast of burden and one obole on a man carrying goods: Bémont, *Rôles Gascons*, 2:125–126, #461. For the repair of the bridge at Moissac the consuls were granted royal permission to levy a pavage both in 1323 and in 1337: Paris, B.N., Doat 127, fol. 145 ff.
35. Boutaric, *Actes du Parlement*, 1:447, #826. M. Quantin, "Histoire de la rivière d'Yonne," *Bull. Yonne* 39 (1885), 442. The king agreed to set aside the judgment in return for 4,000 *livres tournois.*
36. *Inventaire-sommaire arch. dépt. Tarn-et-Garonne*, Ser. A, p. 216.
37. Artière, *Documents*, p. 78, #169.
38. D. Rey, *Etudes archéologiques sur le vieux Millau*, 2: *Le Pont Vieux* (Millau, 1923), pp. 27–29.
39. Artière, *Documents*, p. 83, #180.

belonging to the abbey. Nevertheless, in 1290 she refused to maintain the bridge.[40] At many places inhabitants of the region passed free over the bridge if not carrying on commerce. This was true at Saint-Nicolas-de-Campagnac in 1261, but at Gaillac residents never paid anything, and strangers did so only if transporting merchandise.[41] The same discrimination against non-residents was in effect at two bridges over the Hers near Toulouse, the Pont-de-Negueromieu or St.-Hilaire and the Pont-d'Isalguier—both infeudated by the capitouls of Toulouse in 1282.[42] The toll for a wagon from outside the district was 12 *d. toulousains*, six times as much as for a resident's. If not hauling loads, residents paid a nominal fee only.

The king's concern for bridges was not limited to the granting of extraordinary tolls and taxes to finance construction. The royal government was interested in extending funds for construction and in encouraging it by various means. To rebuild the bridge at Corbeil the royal treasury advanced to the provost of Paris in 1301 the sum of 300 *l. tournois*, the sum to be recovered from tolls on merchandise at Corbeil.[43] On various occasions the king allowed timber to be cut in the royal forests to repair bridges, as at Tours in 1309, and at Carbonne (Haute-Garonne) in 1356.[44] The king donated wood to rebuild the bridge at Agen in 1347.[45]

The royal government also acquired ownership of bridges in this period, as at Millau. In Paris, although the Grand Pont had been donated to the bishop in the ninth century by Charles the Bald, in the thirteenth and fourteenth century the royal government was responsible for repairs. When in 1284 the cathedral chapter ordered some wooden planks to be replaced with a stone arch, the members were careful to obtain a royal charter stating that their rights were in no way changed by this act and that they were in no way obligated for the future.[46]

At Avignon over-zealous royal officials attempted to expropriate the

40. J. Schneider, *La Ville de Metz* (Nancy, 1950), p. 18, n. 52, the bridge over the Orne.
41. E. Germer-Durand, "Le Prieuré et le pont de Saint-Nicolas de Campagnac," *Mém. Acad. Gard* 27 (1863), 185. E. Rossignol, "Droits de navigation dans le Tarn, à Saint Juéry et de passage du pont de Gaillac," *Congrès scientifique de France*, 28e session (1861), 4:580.
42. Archives de la ville de Toulouse, DD 216.
43. Viard, *Les Journaux de Philippe IV*, col. 750.
44. L. de Grandmaison, ed., *Cartulaire de l'archevêché de Tours* (Tours, 1892), 1:304–305. *Ord.*, 3:82.
45. A. Magen, ed., *Jurades de la ville d'Agen (1345–1355)* (Auch, 1894), pp. 125–126.
46. B.E.C. Guérard, ed., *Cartulaire de l'église de Notre-Dame de Paris* (Paris, 1850), 4,2:491–492.

Pont-Saint-Bénézet. The first move was to build a tower in 1293 in Villeneuve-lès-Avignon at the entrance to the Pont-Saint-Bénézet.[47] Then in 1311 the royal agents cut the cords necessary to the operation of the rectors' cable ferry (the Pont-Saint-Bénézet being then impassable) and established their own. On the rectors' complaints the king ordered the seneschal of Beaucaire to let the rectors peacefully enjoy the revenues of their ferry.[48] There was another seizure of the port in 1324, consisting this time in the appropriation of boats belonging to the rectors of the Pont-Saint-Bénézet. The seneschal accused the rectors of neglecting to make repairs, although sometimes in the winter the Rhone was impassable, and even of having broken down the bridge, passage across which was gratuitous, so as to receive tolls from the ferry. At an inquest in 1331 twenty-five witnesses declared that the bridge belonged to the *opus pontis.* In the past whenever the bridge had become unusable, the brothers of the bridge had established a ferry and collected tolls. Some of the witnesses accused the royal officials of destroying one arch of the bridge. One man testified that the rectors had been in control of the bridge for 110 years. (Actually, the transfer from the brothers of the bridge to the rectors of the town must have taken place more nearly seventy years previously). Following the inquest, royal agents restored the bridge to the rectors.[49] The failure of these attempts to expropriate the Pont-Saint-Bénézet did not unduly discourage the royal officials, and they made a number of later attempts to the same end.

At Montauban it was royal determination that was responsible for the monumental brick bridge; the government coerced the town into financing it. There is plenty of evidence of apathy on the part of the citizens. As early as 1144 Alphonse Jourdain, count of Toulouse, who had promoted the construction of the Pont-de-la-Daurade at that city, had provided in a deed that the residents of Montauban should build a bridge. Nothing was heard of the project until the last half of the thirteenth century, when the consuls provided in 1264 and 1275 that certain fines should go toward bridge construction. In particular, anyone violating a sumptuary ordinance should pay a

47. M. Falque, *Le Procès du Rhône et les contestations sur la propriété d'Avignon (1302–1818)* (Paris, 1908), p. 19. In 1335 royal agents seized the bridgehead at Saint-Colombe across from Vienne and built a tower, an action which set the stage for the subsequent acquisition of Dauphiny by the king: P. Fournier, *Le Royaume d'Arles et de Vienne (1138–1378)* (Paris, 1891), pp. 412–415.

48. Falque, *Le Procès*, p. 23.

49. Falque, *Le Procès*, p. 27.

penalty of 1000 bricks for the building of the church and bridge. Again, in 1291 the town bought an island upon which to place bridge piers.[50]

The citizens of Montauban actually built a timber structure, and such a bridge was in use in the second decade of the fourteenth century, and very likely considerably earlier. However, the royal government insisted upon a brick bridge. Only upon the order of Philip IV was work finally started in 1304. The king's document decreed that all strangers were to pay a toll at Montauban, the proceeds to go to the construction of the bridge. In addition, he made a donation to the enterprise, and he commanded the neighboring communities to contribute to the project. There were to be three strong towers on the bridge, one at each end and one in the middle, all three of which the king was to garrison. He put in charge of construction Etienne de Ferrières, royal castellan of Montauban, and Mathieu de Verdun, bourgeois. The consuls, however, refused to contribute to the work. In 1314 the seneschal of Quercy and two judges went to Montauban to inspect the bridge, estimate sums necessary to finish it, and force the consuls to pay the contractors from taxes and other contributions. The following year Etienne de Ferrières and Mathieu de Verdun and four notables concluded that another twenty years would be needed to complete the bridge and that the consuls and their successors should pay the contractors annually 200 *livres tournois.* The sum was to be raised from specified taxes and collected by two *prud'-hommes.* If this proved inadequate, the consuls were to make up the 200 *livres* by a tax on each inhabitant of Montauban.

Just how unwelcome to the townspeople were the activites of the royal officials is shown by the actions of the consuls. They complained of the expense and defended themselves by insisting that they had done more than their share in repairing the timber bridge, damaged in a flood. The consuls unwisely cited Mathieu de Courtes-Jumelles, one of the royal commissioners, before the Parlement of Paris. However, on 9 September 1321 the court decided that the consuls had appropriated for their own use part of the subvention of Philip IV and had shown negligence in building the bridge. Accordingly, the court abolished the consulate of Montauban, took all of its revenues into the king's hand, and determined that the bridge should be finished at the town's expense. The commune was obliged to pay 1000 *livres* damages to Mathieu de Courtes-Jumelles and a fine of 20,000 *livres.* The bridge seems to have been completed about 1335.

50. Didron aîné [A.N. Didron], "Architecture civile du moyen-âge. Le pont de Montauban," *Annales arch.* 16 (1856), 39–48.

The novelty in the history of the bridge at Montauban was not that the royal government authorized tolls and taxes, or that it made a contribution towards construction. The king had been doing this for fifty years. The surprising thing at Montauban is that royal officials were in charge of building the structure and that in effect they bludgeoned the townspeople into erecting a brick bridge, when the latter were content with a timber structure. For several hundred years before this time the inception of such a project had ordinarily been due to local individuals or groups. An excellent example is the building of the Pont-Saint-Esprit at Saint-Saturnin-du-Port, surely not the spot on the Rhone where in the year 1265 a bridge was most needed. Saint-Saturnin was not a point where major roads crossed, and it was only some fifty kilometers north of Avignon, while the next bridge over the Rhone to the north was almost 150 kilometers away. Vienne, for its part, was only some twenty-eight kilometers south of Lyons. It is obvious that central planning would have spaced bridges over the Rhone differently, but that at Saint-Saturnin and at Vienne local people were willing to make the sacrifices necessary to build bridges there.

In the period from 1250 to 1350 it was still almost always true that it was local people who initiated bridgebuilding, but there was an important difference from previous times in the chief reliance for financing. Towns no longer were requesting of rulers permission to solicit alms for bridges, as had been true in the second half of the twelfth century at Agen. What towns like Cahors and Villeneuve-sur-Lot wanted in the second half of the thirteenth century and later was the authorization to levy tolls and taxes, for in this period these were becoming the proved means of financing the construction of bridges.

Government encouragement of bridgebuilding resulted in a rapid increase in the number of spans built in this period. In the Agenais the king of England authorized tolls and ordered the convocation of an assembly to authorize a tax on the region for this purpose, and the kings of France actively promoted bridgebuilding by similar means. Philip the Fair manifested the same drive toward expansion of royal power in bridge construction and maintenance, as he did in his relations with the papacy, the empire, and the nobility. Under his immediate successors royal agents continued his policies. If there was less coercion under the Valois, they continued to encourage bridge construction by granting permission to levy tolls and *barras* and by gifts of timber and money. The result was a high level of bridgebuilding and maintenance from 1250 to 1350.

8

Financing Bridges

A Problem in Maintenance (ca. 1350–ca. 1500)

NO NEW IDEAS on bridges were developed in the last century and a half of the Middle Ages, nor were a significant number of new spans built in this period. The great bridges across the Loire, Garonne, and Rhone, numerous spans across the Seine, and hundreds of bridges elsewhere in France had been built long before the middle of the fourteenth century, and after that the challenge to contemporaries was to keep the bridges passable or to rebuild them in the more enduring material of stone. Two new bridges were constructed at Paris—the Pont-Saint-Michel (1378) and the Pont-Notre-Dame (1413–1416)—but on the whole the period from the middle of the fourteenth century to the end of the fifteenth was not an era favorable to bridgebuilding.

The evidence collected in this study shows that the number of bridges first attested in the fourteenth century dropped to 165 (from 202 in the thirteenth century), a decline that continued into the fifteenth century, when only 96 new bridges are attested. In the fourteenth century most of the decline occurred in northern France, while Dauphiny went against the trend, with 36 new bridges in the century when it came under the rule of the king of France as opposed to only 22 in the thirteenth century. Even in those areas where the number of newly-mentioned bridges held its own in the fourteenth century, there was a decline in the fifteenth century.

Catastrophic mortality due to the Black Death and the attendant economic dislocations inhibited the construction of many new bridges. In addition, the Hundred Years' War was detrimental to bridges in a variety of ways. One was their actual destruction, sometimes by the enemy, as at Vaas (Sarthe), sometimes by the defenders of the countryside, as at Beauvais before 1358 and at Grenade-sur-Garonne (Haute-Garonne) before 1350 by the Toulousains.[1] Then, too, bridge properties were destroyed, and it was

1. F. Legeay, *Recherches historiques sur Vaas et Lavernat* (Paris, 1855), pp. 9, 11, 15. A. de Bourmont, "Les Ponts de Vaas," *Rev. Maine* 22 (1887), 173–174. *Inventaire-sommaire arch. com. Oise. Beauvais*, p. 2. *Inventaire arch. com. Toulouse*, 1:463.

harder for the neighborhood to raise money for taxes and tolls. Much of what was collected went to construction of fortifications rather than to bridge-building, so that the latter was slowed.

The former concept of the bridge as private property lingered on in the fourteenth and fifteenth centuries at some places where the recipient of tolls regarded them as a source of income. There continued to be bridges whose tolls were burdened by payment of sums to individuals, to convents, and to hospitals which made no contribution to maintenance. It is not surprising that ecclesiastical institutions should continue to consider a bridge a pious work, but this idea was shared by many others. At various times and places in the late fourteenth century, as at Romans, substantial sums were collected for bridge repair, and the endowment of the Pont d'Orléans greatly increased between the middle of the fourteenth century and the middle of the fifteenth century. Nevertheless, a public demanding passable bridges was unwilling to await the uncertainties of individual generosity in supplying them. The fifteenth century saw the triumph of the concept of the bridge as a public work with government responsibility for maintenance and the taxation of the residents of the neighborhood for its upkeep.

At this time bridgebuilding commonly was the responsibility of the towns, although sometimes the initiative came from the king or the nobility. Thus the privileges of Figeac (Lot) in 1394, Saint-Geniez (Aveyron) in 1356, Villefranche-en-Périgord in 1357, and Montolieu (Aude) in 1392 listed among the functions of the consuls that of keeping up bridges.[2] At Rouen the bridge, which in the twelfth century had been the responsibility of the dukes of Normandy, was under the care of the commune in the fourteenth.[3] The town in 1375 requested and received from the king permission to collect a *pontenage* on merchandise passing up or down the river under the bridge, because the latter was in such ruinous condition that two arches had fallen into the stream.[4] Tours obtained permission to levy a similar toll in 1373 for the repair of bridges on the Cher and Loire.[5]

Extensive information on bridge finances at the towns of Albi, Agen, Lyons, Orleans, and Romans has survived from the second half of the fourteenth and from the fifteenth century. At Albi, Agen, and Lyons at this time

2. *Ord.*, 7:656; 3:158; 206; 7:500. Montolieu (Aude), cant. and arr. Carcassonne. Villefranche-de-Belvez (Dordogne).
3. *Ord.*, 11:421.
4. E. de Fréville, *Mémoire sur le commerce maritime de Rouen* (Rouen, 1857), 2:144.
5. J.M.A. Delaville-Le-Roulx, ed., *Registres de la ville de Tours* (Tours, 1878–1881), 2:154.

the bridge was directly under the control of the town government; at Romans it was the property of the chapter of St. Bernard and the commune jointly; and at Orleans the bridge with its possessions constituted an *opus pontis* with a large endowment managed by three *proviseurs* appointed by the governor of the duchy of Orleans, the choice being approved by the citizens in an assembly. At none of these places were the tolls alone adequate to finance needed repairs; and at all but Albi donations were solicited, perhaps because the concept of public utility was well-established at Albi long before it was at the others. In the relevant documents the idea of the bridge as a pious work was reaffirmed. In 1352 and 1353 the bishop of Agen promoted the *opus pontis* there by classifying it as a work of charity, so that legacies for good works might be applied to it. An episcopal document of 13 May 1352 stated that certain goods left to Christ's poor or to undetermined pious uses were to be given to the fabric of the bridge or to the repair of roads. The document describes the bridge as worthy of merit and praiseworthy, the most pious among works of charity. One reason for this was that its ruin would mean "the destruction of men."[6] The records of the deliberations of the town government of Agen in 1353 mention legacies turned over by the bishop to the *opus pontis.* One amounted to the large sum of 148 *marcs.*[7]

The general equivalence of the different forms of good deeds, as largesse to the poor and gifts to bridge and road construction, extended also to pilgrimages. The consuls and sworn men of Agen decided on 4 September 1350 to send a deputation to the pope at Avignon to beg him to grant to the inhabitants of Agen the same spiritual benefits as they would have received had they made a pilgrimage to *Rome la velha* (the old Rome), on their donation of the cost of the journey to the building of the bridge over the Garonne at Agen.[8] The town records do not mention the outcome of the mission, but toward the end of the century the equivalence in spiritual merit of bridge-building and pilgrimages was recognized at Romans.

The history of bridge construction at Romans in the fourteenth and fifteenth centuries was typical in that the chapter of St. Bernard divested itself of its responsibility in favor of the commune, and the role of charity dwin-

6. G. Tholin, "Chartes d'Agen se rapportant au règne de Jean le Bon et Charles V," *Arch. Gironde* 34 (1899), 163–164.
7. The records of the deliberations of the town government of Agen mention legacies under the dates of May 28, June 10, and December 16 which were turned over to the *opus pontis* by the bishop: A. Magen, ed., *Jurades de la ville d'Agen* (Auch, 1894), pp. 321, 333.
8. Magen, *Jurades*, p. 212.

dled to negligible proportions. By a royal arbitral sentence of 1360 the obligation to maintain the bridge had been divided equally between the chapter of St. Bernard, which received the tolls, and the citizens of Romans, who passed free over the structure. In 1366 an excise tax on wine was levied for bridge maintenance. It was the intiative of the citizens that resulted in replacing a timber section with a stone arch, beginning in 1387. "Moved by the love of God and of the public good, Guillaume de Sainte-Croix, dit Perrin, merchant of Romans" on March 16 of that year went to the judge of the secular court to propose rebuilding in stone, as it was formerly, the arch next to the chapel of Notre Dame on the bridge over the Isère.[9] Guillaume states that the chapter, the community, and various persons had promised large sums. Accordingly, the judge appointed to be in charge of the project a proctor, a treasurer, a representative of the chapter, and one of the bourgeoisie.

Construction began on 11 March 1388, and the final payment was made to a workman five years later. The expenses of reconstruction amounted to 2797 florins and 6 *gros*, and the cost was met by receipts of 2795 florins and 7 *gros*, with the treasurer donating the missing 23 *gros*. The chapter and the commune each contributed 600 florins, but more than half the receipts came from charity. For example, two men donated a total of 40 florins 10 *gros* as a substitute for pilgrimages to Saint James of Compostella which they had been unable to make. Forty-nine legacies accounted for 769 florins 6 *gros*, charitable solicitations produced 94 florins 6 *gros*, and five alms trunks provided 9 florins, while all the rest was contributed by other individuals.

The reconstruction of one stone arch (1388–1393) exhausted the charitable resources accumulated over several years, but in 1393 the bridge was once again impassable, this time due to the ruinous condition of the fourth arch opposite Romans. Forbidden by the governor of Dauphiny to repair their bridge out of fear that the bands of Raymond de Turenne might enter the town, the citizens voted to levy a *taille* to reconstruct the arch adjacent to the left bank.[10] Four years later, in 1397, the overflowing Isère carried away the arch which had been so recently repaired, and in 1398 the chapter set about levying taxes for its reconstruction. It also established a cable ferry and went into court to defend its right to do so against the officers of the dauphin.

9. U. Chevalier, "Notice historique sur le pont de Romans," *Bull. Drôme* 2 (1867), 315. For the excise tax on wine, see *Inventaire-sommaire arch. dépt. Drôme*, Ser. E, 3:202.

10. P.E. Giraud, *Essai historique sur l'abbaye de S. Bernard et sur la ville de Romans* (Lyons, 1856–1869), 3:341.

Witnesses testified that three times between 1350 and 1397 the bridge had been washed out and that each time it had been the chapter that had set up a ferry.[11]

In the fifteenth century bridge repairs at Romans were paid for by the *commun-du-vin*, a charge of one-sixteenth of the value of wine entering the town. In 1397 the fourth arch towards the Péage de Pisançon (opposite from Romans) and the center of the bridge were repaired, and in 1409 the first arch and the chapel of Notre Dame were reconstructed. When the bridge was impassable, the loss of revenue from the salt merchants detouring around Romans made repairs imperative. In 1426 the chapter authorized a new tax for maintenance, and in 1427 the estates of the province donated 200 florins to the repair of the bridge, but the primary responsibility remained with the town of Romans. In the middle of the century the chapter followed the example of other ecclesiastical bodies and of individuals in relinquishing ownership of the bridge in favor of the town. In 1448 letters patent of the dauphin, agreed to by the archbishop of Vienne and the chapter of Saint Bernard, allowed the inhabitants of Romans to collect the customary charges on strangers on condition of maintaining and reconstructing the bridge.[12]

At Agen the chief financial resources for the bridge in the fourteenth century consisted of the *barragium* and of an excise tax, the *souquet.* Deliberations of the town council of Agen for the years 1345–1355[13] show the effect of the Hundred Years' War on the bridge. The bridge was fortified and ramparts of the town were built in 1345 and 1346. On the bank opposite Agen the town house and all other houses beyond and on the bridge were ordered demolished, routine procedure for a commune expecting to undergo a siege. A stone tower, protected by barriers and a drawbridge, was rendered useless soon after its completion by the disastrous flood of 16 March 1347, which destroyed the middle section of the bridge and left the tower on the farther bank isolated from the town.[14] In time of war even the regular income of the bridge could not be counted on—in March 1349, in February 1350, and in March 1353 the council was obliged to reduce sums owed by farmers of the *barra.* In one case the excuse for the abatement was the presence of troops of the duke of Lancaster.[15] The demands of the enceinte pushed the claims

11. Chevalier, "Notice," p. 318.
12. Chevalier, "Notice," pp. 319–321.
13. Magen, *Jurades*, passim.
14. Ibid., pp. 4, 33, 40, 68, 72, 105.
15. Ibid., pp. 156, 197.

of the bridge somewhat into the background, and on 23 August 1352 the town council decided that the bridge planking should be used to enclose the town.[16] Some months before, on February 9, all notables present at a council meeting had sworn on the Holy Gospels not to divert *denier ni maille* of funds belonging to the bridge, not gifts, legacies, nor fines, except for the *souquet* and the *barrage*, the income from which was mortgaged to repay a loan made the previous year.[17]

Funds for the bridge were chiefly supplied by a *barra* collected at the bridge and four other points in the town, but these sums were supplemented by donations and legacies, and later by the *souquet.* After the flood governmental help was forthcoming: the count of Armagnac gave 500 *l.* and the king 2000 *l.*, half in money and half in timber.[18] The deliberations of 23 September 1351 mentioned "the little bridge of boats given by the king of Navarre" and provided that it should be completed with piles and beams from the *hôtels* of rebels as well as timber belonging to the town, and that people should donate their labor for this purpose.[19] On 23 April 1352 the royal government authorized the collection of a toll of 12 *d. t.* on every barrel of wine passing Agen by land or water, the funds to be used for the construction of the bridge and for no other purpose, "for the advantage of the state and the security of the country."[20] The next day the king extended for two years permission for the inhabitants of Agen to cut wood in the forests of Toulouse and Albi for their walls and their bridge, which they had not been able to do "as well on account of poverty as the pressure of the Gascon war."[21]

Difficulties due to floods, war, and a shortage of funds were compounded by ineffective supervision and incompetent contractors. Finally on 24 February 1353 there was an agreement between the consuls and *operarius*, Pierre Bosiget, on the one hand, and Mestre Raymond Martinola on the other to make a scaffolding and to complete within thirty days a service bridge. Martinola was to receive 16 *l. de petits tournois* for forty days' work.[22] Agen, after six years of putting up with a bridge of boats, finally acquired a timber structure suitable for pedestrians.[23]

16. Ibid., p. 294.
17. Ibid., p. 268.
18. Ibid., pp. 104, 137, for 12 and 27 March 1347.
19. Ibid., pp. 250–251.
20. Tholin, "Chartes," p. 158.
21. Tholin, "Chartes," pp. 159–160.
22. Magen, *Jurades*, p. 307. An *operarius* was in charge of a bridge, responsible for its repair, but he might or might not work on a bridge.
23. On 8 March 1354 toll collectors for the bridge of boats applied for a reduction in

An important event in the bridge's history at Agen was the signing of a contract on 28 December 1381 between the consuls and *prud'hommes* on the one hand and Jean de l'Eglise, bourgeois, merchant, and resident of Agen, on the other. Jean de l'Eglise engaged for 10,000 francs to complete the bridge within ten years for "the necessity, utility, profit, and honor of the said community."[24] The document explained that "because of the force of the flood waters of the Garonne"[25] it had not been possible to keep the bridge passable for long. Jean de l'Eglise promised to support the wooden roadbed with three stone piers and with chestnut piling. The sum of 1,000 francs annually was to be made up by turning over to the contractor the *souquet* on grain, the *souquet* and *arrière-souquet* on wine, the toll on crossing the Garonne, charges on boats going up or down the river, and the *barragium*; and if all these failed to make up the sum, the consuls promised to find the money.[26] Jean de l'Eglise was the contractor the town had been looking for. The bridge was passable in three years and completed in ten. Furthermore, it seems to have endured until the flood of 1435.[27]

The Pont-de-la-Guillotière over the Rhone at Lyons, like the bridge at Agen, suffered from floods in the fourteenth and fifteenth centuries. Both were financed chiefly by tolls and taxes. Unlike Agen, Lyons was struggling in the fourteenth and fifteenth centuries not only to keep its bridge passable but to rebuild it in stone. The municipality from time to time appealed to the king for permission to levy extra charges and to the pope for indulgences. When the commune received the bridge from the abbey of Chassagne in 1334–1335, most of it was in wood. The abutment on the town's side of the river, two arches, and several piers were in stone, substantially as had been the case in 1260. A document of 5 March 1361 in which King John granted a *barragium* stated that "the impetuous velocity of the waters" had so damaged the bridge that in many places it threatened to fall, and that passage of the stream was by boat to the great prejudice of the town and region.[28] The *barragium* was to be one obole for pedestrians and two for horseman, but on 6 September 1364 King Charles V doubled the charges, thus providing funds

their payments to the town on the grounds that a flood had carried off two of their boats and that pedestrians were using the service bridge: Magen, *Jurades*, p. 340.

24. O. Fallières, "Le Pont d'Agen en 1381," *Congrès archéologique*, 68e session (1901), p. 434.
25. Ibid.
26. Fallières, "Le Pont," p. 439, n. 1.
27. G. Tholin, "Les Ponts sur la Garonne," *Rev. Agenais* 5 (1878), 447.
28. M.–C. Guigue, "Notre-Dame de Lyon; recherches sur l'origine du pont de la Guillotière," *Mém. Lyon* (1874–1875), p. 230, n. 2.

to work on the stone bridge. In 1370 a deed extending the *barragium* explained that two-thirds of the bridge was in timber, which entailed expensive maintenance. Moreover the main part of the river, which formerly passed under the stone section, had shifted and now took its course under the timber bridge.[29]

The efforts of the town of Lyons to obtain charitable aid for the Pont-de-la-Guillotière through papal indulgences were exceptionally fruitful. This seems to have been due to the generosity of the indulgence and the organization of a society of bridge benefactors. In May 1384 Pope Clement VII gave a plenary indulgence in *articulo mortis* to those becoming members of the confraternity of the Holy Spirit of the chapel of the bridge over the Rhone, who helped with the repair of the bridge, and who confessed to a priest. The membership fee was set at 6 *gros* annually (16 *gros* were equivalent to one franc). The consuls sent to the dioceses of Bourges, Bordeaux, Vienne, Embrun, Gap, Sisteron, and Puy to publicize the confraternity, and membership grew apace. Before the end of the fourteenth century in Lyons alone there were 4,178 members. In 1368 several arches had already been added, and in 1389 the first bridge portal had been built at the end of the bridge in Guillotière.[30] Before 1407 almost two-thirds of the bridge had been completed in masonry.

Early in the fifteenth century two catastrophic floods damaged the Pont-de-la-Guillotière. The first of these, in 1407, entailed extensive repairs to the arches of La Trappe, of the Cross, and of Saint Nicolas, and one pier and arch needed to be completely rebuilt. For a number of years transportation across the Rhone was by ferry. Construction exhausted the funds of the confraternity, and the commune appealed to the government for aid. On the spiritual level the dukes of Berry and Burgundy successfully appealed to the papacy for indulgences (granted 1410) on the grounds that a large part of the bridge had already fallen into the river and that the bridge, if not repaired, would be an irreparable loss to the realm, as the city was "situated between the rivers Saône and Rhone, key to the realm on the frontiers and marches of the

29. After 1378 a stone arch was begun to join another stone arch with the wooden tower. Apparently this arch was completed by Easter, 1381, and two years later (16 September 1383) the consuls bought a large island "contiguous to the stone pier of the wooden bridge, where the stairs descend to the said island": Guigue, "Notre-Dame de Lyon," pp. 231–233.

30. Guigue, "Notre-Dame de Lyon," p. 238. Pope Clement VII on 11 January 1384 accorded indulgences of 100 days to those who visited the Holy Spirit chapel to be built at the entrance to the bridge and to those who aided it and the bridge with their alms: M.–C. Guigue, *Cartulaire municipal de la ville de Lyon* (Lyons, 1876), p. 186.

empire."[31] On the temporal level, the government on 1 September 1410 authorized the transferral of monies from the fortifications fund to the reconstruction of the bridge. One-fourth part of the tenth on the sale of wine and the tax on grain, which had been applied to the upkeep of buildings and fortifications, was to go to the repair of the town bridge, "because of the urgent necessity of repairs to the said bridge, which is essential to the enceinte."[32] It was estimated that in twenty years a fourth part of the above taxes should amount to about 950 *l. t.*

A sudden flood caused by melting snow in 1414 carried off two of the new arches and some of the old. The Lyonnais requested the grant of a *barragium* on the grounds that their own finances had been severely strained by replacing fallen stone arches with timber spans, by building centerings to support arches in danger of collapse, and by furnishing boats and boatmen to transport men and provisions over the said river. The royal document of 16 September 1418 granting the *barragium* described the bridge as two-thirds stone, one half of which had been rebuilt in the previous twenty-seven years. The length was given as three crossbow shots, while a twentieth-century estimate was 655 meters.[33]

Some evidence on the finances of the Pont-de-la Guillotière is available from the consular registers of the town of Lyons published by M.-C. Guigue. The consuls authorized repairs and reconstruction of collapsed arches and fallen piers, appointed overseers of the work, and approved payment of bills. The chief support was the *barragium* of the bridge, collected outside as well as in Lyons. On 26 September 1416 the consuls decided to "take it into the commune's hand," but on 20 September 1419 we find them complaining that since in each of the last two years the town had collected no more than 548 *l.* 12 *s.* 7 *d. tourn.*, it was better to farm the *barrage* for 600 *l. tourn.* However, two years later in September 1421 a collector was willing to pay 680 *l.* 11 *s.* 6 *d. tourn.* for the privilege. The bridge also owned property, "la grange et le pré du pont," and collected *lods et ventes.*[34] The quest of the bridge over the Rhone, that is, the solicitation of charitable contributions for the bridge, was farmed, as was usual for bridges at this time. There was also an alms trunk opened once a year and found to contain sums ranging from 44 *l.* to more than 80 *l.* Despite the fact that the records refer to the fabric of the bridge,

31. Guigue, "Notre-Dame de Lyon," p. 242.
32. Ibid., p. 246, n. 1.
33. Ibid., pp. 248–249. C.S. Whitney, *Bridges* (New York, 1929), p. 92.
34. M.–C. Guigue, ed., *Registres consulaires de la ville de Lyon* (Lyon, 1882), pp. 347 (22 December 1421), 195 (28 October 1419).

euvre du pont, the consuls made little distinction between the revenues of the bridge and those of the town. Once the consuls ordered the opening of the alms trunk, called "le plot du pardon du pont de Rosne," because money was needed for fortifications and at another time to pay the salary of the seneschal of Lyons.[35]

Unfortunately our information is too fragmentary to give any clear picture of the bridge finances at Lyons, but we know that the people of the commune continued to request the renewal of authorization for the *barragium*; and once again, after a disheartening series of collapsed arches, they appealed to the pope for aid. A plenary indulgence in *articulo mortis* was granted by Eugene IV in 1436 to anyone who himself worked on the bridge for ten days or who paid an equivalent sum for someone else to work on the bridge. The indulgence was successful in bringing in funds, and the list of those qualifying included people from all walks of life.[36] Many lived in places outside Lyons, including Romans and Bourg-en-Bresse. During the rest of the century the Lyonnais vigorously pressed the rebuilding of the bridge in stone, completing it finally in 1572.

Among French bridges in the late fourteenth and fifteenth centuries only the Pont-de-la-Guillotière profited substantially from contributions. Elsewhere, except at Romans in the case of one arch in 1388–1392, the charitable receipts were slender at any one time. Even at Orleans, where the bridge was heavily endowed, annual solicitations did not produce significant amounts. The bridge at Orleans was at least a little older than the Pont-Saint-Bénézet and the Pont-de-la-Guillotière; it had been conceived in the same tradition, and in the fifteenth century it continued to have much of the independence appropriate to a great twelfth-century bridge. A chartulary and some accounts have survived from the bridge at Orleans and the hospital of Saint-Antoine. As a legally constituted body, the bridge and hospital corporation had its own silver seal to affix its impression to its documents.[37] The three masters, governors, or *proviseurs* of the bridge were appointed by the lieutenant governor of the duchy at the request of the proctors, bourgeois, and inhabitants of the town. They were to have the powers of office and to pursue, uphold, and defend the rights, causes, quarrels, and business of the

35. Guigue, *Registres consulaires*, pp. 185 (21 August 1419), 248 (19 June 1420).
36. *Inventaire-sommaire arch. com. Lyon*, 2:187–190. Guigue, "Notre-Dame de Lyon," pp. 250–251.
37. *Inventaire-sommaire arch. com. Orléans*, 1:193. The seal was ordered by the *proviseurs* in 1405–1407.

bridge. They were to receive the revenues and render an accounting at the end of their three years.[38]

The *opus* of the bridge and hospital was responsible for the salaries of the *proviseurs* and staff of the hospital, for hospital expenses, for paying *cens*, for the management of some of the vineyards belonging to the bridge, and for masses in the chapel of Saint-Antoine in memory of donors to the bridge; but these were all minor expenses as compared with the maintenance of the bridge, hospital buildings, and other properties of the *opus*. The *proviseurs* devoted much of their time to keeping the bridge passable, and they supervised routine repairs, such as the driving of piles around the starlings. The accounts mention a fire on the arch of Saint-Antoine in 1397–1398, the cleaning of four piers next the hanging mills in 1402–1404, and the rebuilding of an arch in 1405–1407.[39] The *proviseurs* were responsible for the maintenance of the hospital and chapel of Saint-Antoine; but in the case of houses belonging to the bridge and hospital, the lease determined who was to make repairs. Sometimes the *opus pontis* was obligated to maintain the house, sometimes the tenant, sometimes both jointly.[40] Other duties of the *proviseurs* included defending the rights of the bridge and receiving its rents and revenues. There were legal expenses to maintain the privileges of the bridge, for example, in 1387–1389 against the chapter of Sainte-Croix of Orleans.[41]

The rights of the bridge as to tolls and other sources of revenue are enumerated in the cartulary: *pontenage*, *lançage*, *quêtes*, and others. The *pontenage* was collected only on Saturdays, when there was held on the bridge a market at which bread was sold. For every cartload of bread or hay 2 *d. p.* were due, and from every stall selling bread 1 *d.* For every cartload of straw brought onto the bridge on Saturday, sold or not, 1 *d. p.* was collected, and for every pack animal carrying bread 1 *d. p.* The bridge also had the perquisite of *lançaige* or *lançage et neufvage*, consisting of a charge of 5 *s. p.* on every new merchant the first time he sent goods under the bridge and on every boat on its maiden voyage.[42] The bridge and hospital also had the right to solicit alms consisting of linen, hemp, and all other goods in the Sologne and the whole duchy of Orleans, a perquisite which, according to the cartu-

38. A. Collin, *Le Pont des Tourelles à Orléans* (Orleans, 1895) pp. 468–469.
39. *Inventaire-sommaire arch. com. Orléans*, 1:90, 92–94, 99, 193–194, 197.
40. Archives hospitalières du Loiret, Ser. A III A 1. Archives du Loiret, CC 920, fols. 4v, 12v, 13r.
41. Archives du Loiret, CC 920, fol. 12r.
42. Archives hospitalières du Loiret, Ser. A III A 1, fol. VI^{XX}XIIIIr. Collin, *Le Pont des Tourelles*, p. 470.

lary, should amount to 10 or 12 *l. t.* every three years. The *pontenage*, the *lançage*, and the *quêtes* were all farmed, the first two at auction on St. Andrew's eve every third year. Among the sources of income the document also enumerates the alms placed in the trunk before the statue of Notre Dame in front of the chapel of Saint-Antoine and rent from the Hôtel-Dieu of Orleans for a mill placed under the bridge (8 *s. p.*).[43]

It was not, however, the tolls or *quêtes* that were the chief support of the bridge and hospital. In 1344 the *pontenage* amounted to only 16 *l.*, the *lançage* to 2 *l.* 4 *s.*, and legacies and alms to 9 *l.* 1 *s.* 10 *d.* The remainder of the income of 157 *l.* 15 *s.* 2 *d.* was made up of 30 *l.* 10 *s.* from vineyards, meadows, and other rural properties, 7 *l.* 18 *s.* from *dimes* and all the rest, two-thirds of the total, from rents on houses.[44] A similar situation obtained in the years 1386–1389. The *pontenage* was farmed for 29 *l. p.*, the *lançage* for 1 *l.* 8 *s. p.*, and the *quêtes* for 6 *l.* 4 *d. p.*, while legacies and gifts brought in 15 *l.* 7 *s.* (plus some bedclothes for the hospital), and the alms trunk only 2 *l.* 4 *s.* However, the total income of the bridge and hospital for 1386–1389 amounted to 569 *l.* 10 *s.* 4 *d.*, more than half of which was again derived from the endowment of houses and lands. The property of the bridge seems to have increased between 1344 and 1449. The earlier document identifies thirty-two houses, the manuscript of 1386–1389 mentions twenty-six on which rent was being paid regularly, but the cartulary of 1449 mentions at least forty-seven, some of them in ruins due to neglect or the English siege. Part of the endowment of the bridge had been donated for specific purposes. In 1370 the sum of 100 gold florins of Florence had been willed to the bridge for the purchase of an annual and perpetual rent to be applied to keeping the bridge clean. Accordingly, a house was purchased and used as a dwelling for those who kept the bridge clean.[45] Evidently this arrangement was unsatisfactory, for in 1420 a legacy provided 24 *s. p.* annual and perpetual rent to provide for the sweeping of the bridge four times a year.[46]

Under ordinary circumstances the *opus* of the bridge at Orleans and the hospital of Saint-Antoine with its endowment accumulated over the years was hardly dependent on tolls, and only in an emergency, as when war or flood damaged it, was it necessary to appeal to the government for special taxes and tolls. Such requests were made after the English siege of 1428–1429 and the

43. Archives hospitalières du Loiret, Ser. A III A 1, fol. VI^{XX}XIIIIr-v.
44. Archives hospitalières du Loiret, Ser. A III A 1, fol. VI^{XX}XIIII.
45. Archives hospitalières du Loiret, Ser. A III A 1, fol. IIII^{XX}v ff. Collin, *Le Pont des Tourelles*, p. 485.
46. Collin, *Le Pont des Tourelles*, p. 485.

floods of 1435, while the bridge fortifications were the responsibility of the town. In the fifteenth century the bridge at Orleans was something of an anomaly, a survival of twelfth-century ideals. The free passage across the bridge (except on Saturdays), the extensive property, and the degree of autonomy were all survivors from a period 300 years earlier. The bridge at Agen, the Pont-Saint-Bénézet at Avignon, and the Pont-de-la-Guillotière had all been conceived in this tradition, but at Agen charity had never provided the bridge with sufficient property to build and maintain it, at Avignon the bridge had been destroyed in the wars, and the Pont-Saint-Bénézet and the Pont-de-la-Guillotière in the thirteenth and early fourteenth centuries had lost at least part of their endowment and passed under the direct control of the municipal government. The bridges at Agen and Lyons then became dependent on taxes and tolls specially authorized by the royal government. At Avignon aid for the bridge was sought from the papacy, which had purchased the city and the county Venaissin in 1348. In the fourteenth century the popes frequently donated large sums to rebuild the bridge, and Clement VI (1342–1353) reconstructed four arches in stone.[47]

In the fourteenth and fifteenth centuries bridges were usually administered by towns supervised by the nobility or by the royal government. At Toulouse the *capitouls* (aldermen) frequently appealed to the Parlement of that city to authorize the financing of bridge repairs. Floods damaged the Pont de Tounis and the Pont-de-la-Daurade in 1413 and the Old Bridge in 1484.[48] The Pont-de-la-Daurade was in such poor repair in the middle of the century that carts were forbidden to cross the structure, and their seizure was authorized by Parlement on 22 August 1454.[49] The cooperation of Parlement was essential to procure funds for the bridge. The court ordered payment of various sums from the *taille* and from the tax on wine for work on the Pont-de-la-Daurade and provided for inspection of repairs to the bridge.[50]

In the fifteenth century the ultimate responsibility for French bridges was recognized as resting with dukes and kings. René, duke of Anjou and titular king of Sicily, promoted bridge construction through grants, permission to levy special taxes, and through authorization to reconstruct bridges. For ex-

47. S. Baluzius, *Vitae paparum Avenionensium*, ed. G. Mollat (Paris, 1914), 1:287.
48. J. Chalande, "Les Formations alluviales dans le bassin de la Garonne à Toulouse depuis le douzième siècle," *Mém. Acad. Toulouse*, 10th ser., 12 (1912), 71.
49. Archives dépt. Haute-Garonne, B 1, fols. 7, 244.
50. Archives dépt. Haute-Garonne, B 4, fol. 355. See also B 1, fol. 226; B 2, fols. 163, 169; B 3, fol. 21; B 4, fols. 264, 309, 369; B 6, fol. 21. Toulouse, Archives de la ville, DD211.

ample, although it was the monks of Saumur who were obligated to maintain the bridge there, after the disastrous flood of 1456 René donated his receipts from mills under the bridge to aid "the public affairs of our country"[51] by rebuilding the fallen arches. To reconstruct bridges he permitted the inhabitants of neighboring parishes to tax themselves, as for example at Motte-de-Bourbon near Loudun (Vienne) in 1455.[52] An unusual form of bridge financing was that at Montouron (Maine-et-Loire), parish of Mazé on the Authion, when René let an *arpent* of land on condition that the lessee should pay 12 *d.* of *cens* and build and maintain a stone arch 3.24 meters wide to replace a timber one.[53] The courts of Anjou compelled negligent owners of bridges to repair them. Because the abbey of Bellefontaine had refused to maintain the bridge at Chaudefonds on the Layon near Chalonnes (Maine-et-Loire), its temporalities were ordered seized by the assizes of Angers in 1451.[54]

One of the features characteristic of the history of French bridges in the fifteenth century was a new emphasis on regular maintenance. The Ponts-de-Cé at the opening of the fifteenth century consisted of three main timber bridges between islands on the Loire; and the records of Yolande of Aragon, duchess of Anjou, and of her son René, show them actively pressing for repairs and conversion to stone. Various tolls and taxes supported repairs to the Ponts-de-Cé: sums from the farm of the "imposition foraine," from the *trespas de Loire*, from charges on boats, and from tolls on wine sold in Angers.[55] After the flood of 1458, because of the urgency of work on the bridge over the Loire, "where no one any longer crossed the Ponts-de-Cé, either on foot or horseback," the king of Sicily gave 500 *l.* for repairs.[56] The Ponts-de-Cé received methodical and continuing maintenance. A carpenter was hired at yearly wages to keep the bridge in repair, and a sergeant supervised the

51. A. Lecoy de la Marche, ed., *Extraits des comptes et mémoriaux du roi René* (Paris, 1873), p. 152.
52. Ibid.
53. Ibid., p. 153. See also Lecoy de la Marche, *Le Roi René* (Paris, 1875), 2:63. Mazé (Maine-et-Loire), arr. Baugé, cant. Beaufort. Another example of ownership of property entailing obligations occurred at St.-Flour (Cantal). In 1250 a bourgeois wishing to acquire a site for his dwelling in the suburb of Pont next to the house of the confraternity of Sanctus Spiritus de Ponte and adjacent to the bridge was granted the property on the condition that he and his successors were to provide the portal built on the bridge with planks, iron, locks, and keys as needed: M. Boudet, ed., *Cartulaire du Prieuré de Saint-Flour* (Monaco, 1910), p. ccxxviii, #XXVI; pp. 64–65.
54. Lecoy de la Marche, *Extraits*, pp. 149–150.
55. Ibid., p. 161 (7 January 1478).
56. Ibid., p. 155.

masons, carpenters, and laborers.[57] There was also a *garde des Ponts-de-Cé.*[58] From time to time there were inspections of the Ponts-de-Cé, such as that made March 31, 1455, by two councillors and auditors of the *Chambre des Comptes* of the king of Sicily, the *receveur ordinaire* of Anjou, the *contre-rolle du pont de See*, a clerk of the *Chambre des Comptes*, five carpenters, and others.[59]

The pattern of governmental activity on the part of the dukes of Anjou vis-à-vis bridge maintenance was that long followed by the royal government. The king appointed officials to inspect bridges, for example, around Paris at Charenton, St. Cloud, and Poissy.[60] The royal government gave donations to bridge construction, and courts provided penalties for failure to maintain bridges. Thus in 1387–1389 a 500 *l.* fine was levied by the assizes of Beaugency on the count of Blois for neglecting to repair the bridge at Millançay (Loiret).[61] When the royal seneschal attempted in 1382 to force the abbey of Notre-Dame-de-Vaas to repair the bridge at Vaas (Sarthe), the convent pleaded both poverty and an absence of responsibility. The seneschal was then ordered to lay an aid on the habitants of the parishes adjacent to the bridge.[62] King Charles V used a variety of tolls and taxes to finance repairs to the bridge and fort of the church of Mantes in 1366–1374. The tolls, levied on traffic by water, could not have been expected to meet the cost, even had they not been burdened with fixed charges: 100 *l.* to two individuals and an annual rent in salt to the Hôtel-Dieu. Accordingly, taxes from two sources were used for the bridge and fortress of Mantes. In 1366 a tax was levied on the people of the countryside surrounding the fortress and bridge of Mantes, and in 1374 Charles V ordered transferred from the aids ordained for the war 200 *l.* every month for five months to pay for the work on the bridge, church, and fortifications at Mantes.[63]

57. Ibid., pp. 159–160.
58. Ibid., pp. 160–161 (9 May 1477).
59. Ibid., p. 151.
60. L. Dessalles, "Rançon du Roi Jean," *Mélanges de littérature et d'histoire* (Paris, 1850), p. 274. For other examples of inspections of bridges, as well as roads, see *Ord.*, 6:278 (1377); 7:329–330 (1389).
61. Archives du Loiret, A 1979, fol. XLIXv. The assizes suspended the fine on the count's promise to repair the bridge, which was in such a state that "one could neither pass nor repass." For a similar fine, see also Archives du Loiret, A 1979, fol. Lr.
62. The aid was not to establish a precedent nor to turn to the disadvantage of the residents: de Bourmont, "Les Ponts de Vaas," pp. 173–174.
63. The parishes and towns of Dangu (Eure), Gisancourt, and Vesly were to pay taxes for the support of the bridge at Mantes, but those at Vernonnet (Eure) and Triel

The king of France was responsible for the maintenance of the bridges of Paris in the fourteenth century and at the beginning of the fifteenth. Charles V initiated the construction of the Pont-Saint-Michel in 1378. Two royal comissioners called together, for consultation on the new bridge, members of Parlement and of the chapter of Notre Dame, the provost of Paris and five bourgeois.[64] The provost of Paris was in charge of the bridge and he forced vagabonds to work on it. The stone structure was completed in 1387.

The gloomy picture painted in the chronicles of the fratricidal Armagnac-Burgundian struggles affected the bridges outside but not in Paris. There the damage was due to floods. At the end of the fourteenth century the Pont-Saint-Michel required repairs and the Petit Pont reconstruction. Accounts of the payer of the town works give the cost of rebuilding the Petit Pont as 21,789 *l.* 3 *s.* 10 *d. p.*, including sums from a fine and from the aids ordained for the works, fortifications, and other needs of the town of Paris. Operations were begun in June 1394 and completed eleven and a half years later.[65]

Shortly after the Petit Pont had been completed, it was washed out once again, on Saint Martin's day (11 November 1406), and on 31 January 1407 the Pont-Saint-Michel suffered the same fate. In an assembly held by the Parlement of Paris, the *Chambre des Comptes*, and the treasurers of France, it was decided that the treasurers and the Hôtel de Ville should each furnish 1,000 *l. p.*, the Parlement 500, the Châtelet a sum to be determined, and the bishop of Paris and others receiving income from the bridge such amounts as they might be able. Work to rebuild the Petit Pont was to start forthwith. The provost of Paris and the provost of the merchants were to appeal for funds to the queen, the king of Sicily, the duke of Berry, and other great men of the realm. (At this time, the king, Charles the Mad, was incapable of governing.) When work was stopped for lack of money, it was concluded that fines, too, should be applied to the bridge repairs.[66] The work on the stone Petit Pont was finished on 10 September 1409, and this appeared to the royal government to be a good opportunity to divest itself of responsibility for the structure. In line with the trend all over France to make bridges the property of

---

were not. The people of Triel had been contributing to the bridge at Poissy: L. Delisle, ed., *Mandements et actes divers de Charles V (1364–1380)* (Paris, 1874), pp. 17, 114, 154, 180, 219, 566, 567, 579, 759. Gisancourt and Vesley (Eure), cant. Gisors. Triel (Seine-et-Oise).

64. H. Sauval, *Histoire et recherches des antiquités de la ville de Paris* (Paris, 1724), 1:225–226.

65. Ibid., 1:217.

66. Ibid., 1:218.

communes, Charles VI added the Petit Pont to the domain of the town of Paris.[67]

It was the town of Paris which built the Pont-Notre-Dame on the site of the Planches de Mibray, on land belonging to the monks of Saint-Magloire. Permission was granted by the convent in 1412, and in 1413 the provost of merchants and the aldermen decided to build the bridge of wood. The king donated timber from his forests of Lyons, Cuisse, and Pontcourt, as well as one-third of the duties levied by the royal government in the town, amounting to more than 35,000 francs, but on condition that the town of Paris was to make all repairs.[68] The king permitted the town to build houses on the bridge, while he retained rights of high, low, and middle justice and a *cens* on the bridge.

The achievement of the last half of the fourteenth century and the fifteenth century in the history of bridges was not so much in building new structures or in replacing timber spans with stone as in the definitive recognition of the bridge as public property. Consequently the government was obligated to see to it that the bridge was passable, and the inhabitants of the region who used the bridge were to be taxed to this purpose. The general tendency of the late fourteenth and of the fifteenth century was to recognize the need for methodical maintenance of bridges. The neglect of the Pont-Notre-Dame by the municipal authorities provoked the bitter anger of the Parisians. In 1440 the bridge was reported as in ruinous condition, but nothing was done. A flood of 1497 damaged it, and the next year the masters of the works asked that the piles be replaced. Before this recommendation could be acted upon, a carpenter reported to the *lieutenant criminel* that the bridge would collapse in a matter of hours. The inhabitants were ordered to leave but not all did so. The fate of the bridge was reported by Corrozet (2nd ed., 1561):

> The year 1499, Friday before All Saints, the twenty-fifth day of October, the Pont-Notre-Dame, supported on piles with sixty houses erected thereon in very seemly order and of the same height, one hour before noon fell into the river Seine eighty-two years after having been built.[69]

67. A. Tuetey, ed., *Journal d'un bourgeois de Paris (1405–1449)* (Paris, 1881), pp. 300–331.
68. Sauval, *Histoire*, 1:228–229. J. de Saint-Victor, *Tableau historique et pittoresque de Paris* (Paris, 1808–1811), 1:174.
69. G. Corrozet, *Les Antiquitez, chroniques, et singularitez de Paris*, 2nd ed. (Paris, 1561), p. 149r.

Four or five persons were killed in the fall of the Pont-Notre-Dame. Public indignation was such that the authorities feared a riot. The provost of the merchants and aldermen of Paris were arrested, imprisoned, and heavily fined, because they were held responsible for the loss of life and property entailed by the fall of the Pont-Notre-Dame.

No such outcry had greeted the collapse of the Old Bridge at Toulouse 300 years earlier, although on that occasion 200 persons were drowned. (A large crowd had gathered on the bridge to watch the ceremony of bathing the cross in May 1281).[70] The tragedy does not seem to have been considered the fault of anyone in particular. It should be remembered that in the thirteenth century bridges were still often considered private property, but in the fifteenth they were usually public works, for which responsibility was fixed. By the end of the fifteenth century the people of France expected and demanded that bridge maintenance be such as to ensure the public's convenience and the safety of lives and property.

70. A.L.C.A. Du Mège, *Histoire des institutions religieuses, politiques, judiciaires et littéraires de la ville de Toulouse* (Toulouse, 1844–1846), 4:232.

9

The Appearance of Medieval Bridges (ca. 1250–ca. 1500)

SOME OF THE MOST IMPRESSIVE French medieval bridges were built between the middle of the thirteenth century and the end of the fifteenth century—among them the Pont-Saint-Esprit and the bridges at Montauban and Cahors. In addition, earlier spans like the bridge at Albi and the Pont-Saint-Bénézet were extensively rebuilt in this period. But many of the structures that should be included in a description of later medieval bridges no longer survive. Nothing that can be identified with certainty as medieval remains of the Pont-de-la-Guillotière,[1] and the immense bridges across the Loire have all disappeared except for half of the bridge at Beaugency. At Paris today the oldest extant bridge, the Pont-Notre-Dame, dates only from the early sixteenth century, and it has been drastically altered from its original form in the interests of navigation. Those bridges that have survived have usually been modernized, and in most cases a study of the extensive documentation is necessary to visualize their appearance in the Middle Ages.

The silhouette of French bridges does not seem to have been altered appreciably in the years 1250–1500. Improvements came in the form of technical advances, notably in the methods of founding piers, while the overall design remained the same. French bridges continued to display ogival, semicircular, and segmental arches, and to support towers, houses, and workshops; mills still clustered around them.

The four remaining arches of the Pont-Saint-Bénézet probably date from the fourteenth century. One can understand the admiration of contemporaries for the Pont-Saint-Bénézet, which at one time had twenty-two arches and spanned some 900 meters. The bridge made an angle about a third of the way across the river from Avignon, evidently so as to cross the island of Barthelasse (Figs. 7, 8, 9). The chapel of St. Nicholas was built on the second pier,

1. M.–C. Guigue, "Notre-Dame de Lyon," *Mém. Lyon* (1874–1875), p. 270.

and in it the tomb of St. Bénézet was placed. The Pont-Saint-Bénézet underwent more vicissitudes than most medieval bridges. In the first place, the current of the river at Avignon is very swift. Moreover, the piers were not well founded. Again and again portions of the bridge collapsed, making it impassable for longer or shorter periods. Already during the period of construction there was trouble caused by the undermining of piers by the river, as we learn from the beatification proceedings for Saint Bénézet. The witness was not interested in the techniques of bridge construction. What impressed him was that when Bénézet was in Burgundy on a fund-raising expedition he announced that they must return to Avignon: he knew that the devil was scattering the stones of a pier at the bridge.[2]

Then, too, the political situation of the commune of Avignon on the edge of the kingdom of France was such as to involve the bridge in quarrels, and even in war. The thirteenth century saw the destruction and rebuilding of the bridge, apparently mostly in timber; in the fourteenth century Popes John XXII and Clement VI repaired the wooden structure; and in the years 1342–1353 the latter rebuilt four arches in stone.[3] Perhaps these are the ones now standing, although subsequently several arches of the bridge were destroyed in the troubles connected with the schism. The struggle to rebuild the entire Pont-Saint-Bénézet in stone continued during the fifteenth century, but in the seventeenth the will to maintain the bridge failed. In 1602 one arch fell, and in 1605 three more; but wooden replacements made the bridge passable, and Louis XIV crossed it in 1660. In 1669 a flood carried away two more arches; in 1674 the remains of St. Bénézet were transferred to the church of the Celestins in Avignon, and in 1715 the chapel on the pier was abandoned. It was by then dangerous to go up on the bridge, and it was only in the nineteenth century that repairs made certain that the four arches now to be seen would not disappear in their turn.

The remnants of the Pont-Saint-Bénézet, amounting to less than a fifth of the old bridge, were obviously much influenced by the ancient Roman bridges of southern France. The semicircular bays through the upper part of the piers resemble those of the bridge at Sommières, and the narrowness of the distance between the voussoirs and the roadbed was typical of the ancient bridges. The large stones of the voussoirs not only resemble Roman stones—they are Roman. There was a French proverb to the effect that a stone house destroyed is a house half rebuilt, meaning that at least the materials are at

2. *Acta sanctorum*, April, vol. 2, new ed. by J. Carnadet (Paris, 1865), p. 268.
3. S. Baluzius, *Vitae paparum Avenionsium*, ed. G. Mollat (Paris, 1914), 1:287.

hand. The medieval successors of the Gallo-Roman inhabitants made use of ancient stones for construction. The manner in which stones of the voussoir are laid up is the same as that of the Pont-du-Gard—there are four parallel courses of stone, generally unconnected with each other, and they are placed without obvious mortar (Figs. 6, 7). The upper portions of the bridge, nevertheless, are made up of smaller stones, apparently because the ancient supply of large ones was exhausted.

Three points which distinguish the Pont-Saint-Bénézet from Roman bridges are the segmental arches, the cutwaters pointed both upstream and down, and the chapel built on the second pier. It has been suggested that the original arches were semicircular, but in the extant bridge they are segmental with arch openings of 31.8–34.8 meters, the width of the piers being about a quarter of this distance. Documentary and architectural evidence combine to show that the bridge was first built at the end of the twelfth century. The lower chapel was built in the Romanesque style, probably on a level with the original bridge. Unfortunately, it proved to be below flood level, and the floor was actually covered by water in 1840 and 1856 to the depth of 0.60 meters.[4] Here was another example of the medieval tendency to build low bridges. At Avignon it was necessary to raise both the bridge and the chapel, especially to place the body of St. Bénézet in safety. At present a staircase takes the visitor down from the roadbed of the bridge to the lower chapel with its Romanesque vaulting; the upper chapel is Gothic in style. It has been suggested that the lateral walls of the chapel are twelfth century and that a floor was inserted in the thirteenth; what we know is that there were repairs after 1400, that the chapel was reconsecrated in 1411, and that the upper apse was built in 1513.[5] Noteworthy in any consideration of the bridge are its narrowness, only 4.7 meters, and the absence of a parapet to protect persons crossing. It is no wonder that when the Grande Mademoiselle, cousin of Louis XIV, came to Avignon in 1660, she wrote,

> I saw the bridge and the Rhone by moonlight; both the one and the other appeared to me very beautiful and frightened me very much, because the

4. A. Sagnier, "Le Pont-Saint-Bénézet," *Congrès arch.*, 49e session (Avignon, 1882), pp. 279–280, and also p. 261. For descriptions of the bridge, see W. Emerson and G. Gromort, *Old Bridges of France* (New York, 1925), p. 42; P. Séjourné, *Grandes Voûtes* (Bourges, 1913–1916), 2:34, n. S''_7; E. Degrand, *Construction* (Paris, 1888), 2:55. The proportions of the rise to the span of the segmental arches are 1/2.86, 1/2.81, and 1/2.82 (Séjourné, *Grand Voûtes*, 2:25, n. 3).
5. L.H. Labande, *Guide archéologique du Congrès d'Avignon*, 76e session, 1 (Paris, 1910), 51.

> Rhone is very swift and very wide and the bridge is very narrow, very high, and in bad repair.[6] (Fig. 7).

The Rhone had a sinister reputation, and Victor Hugo referred to it as a tiger.

Only fifty kilometers up the Rhone from Avignon is the Pont-Saint-Esprit, from the engineering point of view a much finer achievement than the Pont-Saint-Bénézet. Begun in 1265, it is the latest in date of the great stone bridges completed in the Middle Ages, and the only one now surviving almost intact (Fig. 10). (This does not count the Pont-de-la-Guillotière at Lyons, which was not finished during the medieval period and which has been substantially altered since that time.) The initiative for the Pont-Saint-Esprit came from the inhabitants of the little town of Saint-Saturnin-du-Port, an allod belonging to the Cluniac priory of Saint Pierre. The citizens obtained permission of the prior to begin work, which they wished to do on August 16, since they had assembled "the necessary stones and lime and sand and boats and all things essential to the beginning of the work, and especially since at the present time the Rhone river is small and suited to working on the said bridge."[7]

Construction of the Pont-Saint-Esprit occupied some forty-four years, as compared with the eleven years for the Pont-Saint-Bénézet. The site was difficult, the river bed being extremely wide with devastating floods at some times and at others the river divided into streams flowing in an expanse of sand. Once completed, nevertheless, the Pont-Saint-Esprit had an enviable record of continuous service very different from that of the bridge at Avignon. In part this was due to superior construction and in part to adequate finances, resulting principally from the donation by King Philip VI in 1328 of a toll on salt for this purpose. The Pont-Saint-Esprit continued to receive good care. Here, as at Avignon, the bridge was understood to be for pedestrians and horses—up to 1774 wheeled vehicles were forbidden, and goods were taken across either on pack horses or by sledge pulled by men.

The bridge at Pont-Saint-Esprit is 740 meters long, with nineteen arches between the embankments, and it makes a slight angle, upstream.[8] A comparison with the Pont-Saint-Bénézet shows definite resemblences. Both bridges have segmental arches, cutwaters pointed upstream and down, and bays piercing the piers, and the voussoirs in each case are formed of four par-

6. J. Girard, *Evocation du vieil Avignon* (Paris, 1958), p. 352.
7. L. Bruguier-Roure, *Cartulaire Pont-Saint-Esprit*, *Mém. Acad. Nîmes*, Ser. 7, vol. 12 (1889), Annexe, pp. 6–7.
8. F. de Dartein, *Etudes sur les ponts en pierre*, 1 (Paris, 1912), 19.

allel courses with little interlocking. On the other hand, the arches of the Pont-Saint-Bénézet, presumably of the fourteenth century and later than the Pont-Saint-Esprit, do not seem to have been copied from the earlier bridge but were rather influenced by the Pont-du-Gard and Roman bridges, such as that at Sommières. The shape of the arches is not the same in the two bridges. Those of the Pont-Saint-Bénézet are more flattened, and the distance between the roadbed and keystones, very small at Avignon, is greater at the Pont-Saint-Esprit. The laying of the roadbed almost on the voussoir was a Roman practice, while the medieval custom was to permit the roadbed to rise substantially above the keystone of the arch. The cutwaters at Pont-Saint-Esprit were very low and only raised to the level of the bays through the tympans in the eighteenth century, when, too, the parapets were raised.

At Pont-Saint-Esprit the largest arch opening measures 35.2 meters across and the piles from 9 to 11 meters. The arches are segmental, but near enough to a semicircle to offer the same disadvantages to navigation. In 1853, at a time when railroads had not yet been built in this region, high water prevented steamboats carrying grain from passing under the Pont-Saint-Esprit. The result was a food shortage at Lyons. Accordingly, the first two medieval arches, counting from the right bank, were removed, and a single large one, 62 meters wide, replaced them, so that steamboats could ascend the Rhone at any time. M. Aymard, the engineer who carried out the modernization, published an account of it in the *Annales des Ponts et Chaussées* for 1859.[9] Removing the voussoirs posed a problem, for they were composed of four parallel bands except that at every fifth voussoir there were three larger stones instead of four. M. Aymard was prevented from removing the stones several at a time, because the mortar was powdery. Some years later chemical analysis of the mortar of the Pont-Saint-Esprit showed it to be an ordinary sand-lime mortar.[10] M. Aymard destroyed the pier between the arches with mines and used cranes to remove the stones. He discovered that the piers towards the right bank were founded on rock; tradition says that those toward the left bank were built on piles. Above the bridge, said M. Aymard,[11] the river was an archipelago. The principal branch of the river, touching the

9. M. Aymard, "Notices sur les travaux exécutés au pont Saint-Esprit," *Annales des Ponts et Chausées*, 3me sér. (1859), pp. 1–48.
10. Bruguier-Roure, *Cartulaire*, *Mém. Acad. Nîmes*, Ser. 7, vol. 17 (1894), Annexe, p. xxix, n. 2.
11. Aymard, "Notice," p. 2. See also Séjourné, *Grandes Voûtes*, 2:25, n. 5, where he gives the *surbaissement*, that is, the proportion of the rise to the span of the largest arch, as 1/3.45, less than a semicircle.

right bank, ran parallel to the bridge before passing under an arch next the left bank.

During World War II the bridge was bombed. Afterwards much of the water from this part of the Rhone was diverted to a ship canal, and since the Pont-Saint-Esprit no longer influenced navigation, the nineteenth-century steel arch was removed and two masonry arches similar to the thirteenth-century arches were substituted.

The reputation of the Pont-Saint Esprit has declined appreciably in the last 200 years. On the one hand, it is hard to sympathize with the eighteenth-century guidebook which states that one cannot sufficiently admire the delicacy of the piers;[12] on the other, in an age of suspension bridges it no longer appears as an important engineering triumph. Also, it has lost both its chapel of Saint Nicolas on the thirteenth pier and the fortifications at both ends of the bridge. F. de Dartein,[13] writing in 1912, was critical of the Pont-Saint-Esprit on artistic grounds. The flattened segmental arches with the keystone only 15 meters above mean water level were displeasing to him. Although he admitted that the Pont-Saint-Esprit was much longer than the bridge at Montauban (Fig. 11), which measures only 20 meters between banks and has only seven ogival arches, he gave the palm to the latter. He preferred the elevation of the bridge at Montauban, 20 meters above water level, as having a stature both elegant and sturdy, which gave it an incontestable artistic superiority. It is, Dartein says, the most monumental of French medieval bridges.[14]

The bridge at Montauban was probably completed about 1335. It was built of brick with very little stone, and the middle of three towers (all now gone) contained a chapel dedicated to Saint Catherine. In 1828 the parapets were removed so as to lay sidewalks and widen the bridge. This occasioned the use of corbelling. At this time, too, the chapel was demolished and a staircase through the pier removed. A view published in 1833 shows one pointed cutwater, flat on top, rising to a level above the keystone.[15]

The Montauban bridge incorporated ideas used at the bridges of Sommières, Béziers, Avignon, and Pont-Saint-Esprit. Notable among these was the inclusion of a bay through the pier, although there the apertures form pointed arches, not rounded arches as at the others. The pointed cutwaters,

12. M.L.R. [C.–M. Saugrain], *Nouveau Voyage de France, géographique, historique et curieux* (Paris, 1771), p. 64.
13. Dartein, *Etudes*, 1:19.
14. Ibid.
15. C. Nodier, J. Taylor, A. Cailleux, *Voyages pittoresque et romantiques dans l'ancienne France* (Paris, 1820–1878), 2:P1. 10.

flat on top, rise to a level just below the bays, as is true at Sommières and Béziers; but unlike them, and perhaps in imitation of the Pont-Saint-Esprit, at Montauban the cutwaters are pointed downstream as well as upstream. Finally, while the bridge at Montauban has many of the features of the bridges situated in southeastern France, it nevertheless has the Gothic arches of the bridges at Cahors on the Lot and at Albi on the Tarn.

The story of the old bridge at Albi, planned ca. 1037, is typical of the history of extant medieval bridges (Figs. 12, 13). At least by the middle of the fourteenth century the structure was of stone, to which repairs were made in brick and stone. Between 1403 and 1410 three arches were rebuilt. In the early fifteenth century there were an iron cross, entrance gates, a drawbridge, a toll house, residences, and defensive towers, as well as gargoyles to carry off the rainwater from the roadbed.[16] All these except the iron cross have long since disappeared. The result is that the medieval bridge at Albi can satisfy neither the romanticists, with its lack of a picturesque air, nor the architects, because of an absence of symmetry. Degrand complained that the bridge at Albi has pointed arches of varying width and piers badly aligned.[17] Even the historians are dissatisfied that the bridge lacks a truly medieval air because the superstructure is wanting, and the bridge has been widened by the addition of brick.

Perhaps the French bridge still extant which best answers to our ideas of what a medieval bridge should have been like is that at Cahors (Fig. 14). Begun in 1308, the Pont Valentré was contemporary with the bridge at Montauban. The third bridge built at Cahors, it was outside the town. It has the typically medieval features of pointed arches, the roadbed placed well above the keystones, cutwaters pointed upstream and square downstream which rise perpendicular to the parapets, and a slight rise in the level of the bridge towards the middle. Most unusual, the bridge retains its three towers. Alterations seem to have been minor. A picture published in 1835 shows a low passageway with a tile roof in front of the tower on the Cahors end of the bridge and also a dormer window since removed.[18]

Only the records, however, can recreate the appearance of the famous bridges across the Loire and Seine or detail the variety of structures medieval bridges supported. These included dwellings, workshops, shops, towers, chapels, toll houses, and even an occasional recluse's cell. For example, in

16. A. Vidal, ed., *Douze Comptes consulaires d'Albi du XIVe siècle* (Albi, 1911), 2:231.
17. Degrand, *Construction*, 2:64–65.
18. Nodier, Taylor, Cailleux, *Voyages*, part 2, vol. 2, P1. 71 bis.

1422 the Pont-Notre-Dame at Paris was described as supporting sixty-four houses belonging to the city and eighteen belonging to other persons, and five were under construction.[19] To a large extent buildings on bridges were occupied by persons having a business interest in the location, as the money changers and jewelers on the Grand Pont at Paris. At Orleans the bridge receipts for 1428–1431 list seven houses "on" the bridge, two of them occupied by ropemakers, one by a baker, and one by a haberdasher.[20] Four of them seem to have been on piers, and three (including the baker's) on the sandy island next the chapel and almonry, where there was a market every Saturday. Plans for rebuilding the bridge after the flood of 1435 specified that one of the piers was to be 5.85 meters wide with a cutwater 5.85 meters long, evidently an area large enough for a house. The plans drawn up in 1437 insisted that the piers should be placed where they had formerly been, so that houses could be built on them for the profit of the bridge. Five houses had been swept away by the flood, and five were rebuilt.[21] The disappearance in the river of five houses seems to have discouraged neither the *proviseurs* of the bridge and hospital nor the tenants, and by 1446 the new houses had been rented.

There were workshops on the Pont-de-Saône at Lyons, too, but here the attitude of the syndics towards the desirability of buildings on bridges was ambivalent. Instead of viewing them as sources of income or prosperity, the syndics of Lyons worried lest such buildings might damage the bridge or encroach on the roadway. In 1309 the town government gave permission to a knight who owned several workshops on the bridge over the Saône to build two stone pillars to support the upper stories of a building on the pier of the Miraculous Arch, but with the proviso that he was not to build anything wider or longer than had been there beforehand; and he was to be held responsible for maintaining the pier and the arch.[22] The same building was burned at a later date, and in 1340 the syndics again gave a grudging consent to its rebuilding for fear of "damage, deterioration, and the cutting of the bridge" in the future.[23] Only when the owner promised to repair at his own expense any such damage was he permitted to rebuild in the area which

19. A.J.V. Le Roux de Lincy, L.M. Tisserand, eds., *Paris et ses historiens aux XIVe et XVe siècles* (Paris, 1867), p. 160.
20. Archives du Loiret, Orleans, CC 930, fols. 7v–8r.
21. Archives hospitalières du Loiret, Ser. A. III A 1, fol. VI^{XX}Xv and fol. VI^{XX}XIr.
22. M.–C. Guigue, *Cartulaire municipal de la ville de Lyon* (Lyons, 1876), pp. 130–131.
23. Ibid., pp. 449–451.

stretched from a mercer's workshop "to the corner of the great wall of the Miraculous Arch."[24]

It was natural that there should have been toll houses on bridges. There was one on a pier at Albi in the early fifteenth century; and at Millau in 1338 an assemblage of royal officers and local dignitaries decided, among other things, that there should be built on the Pont Vieux an office for the toll keeper with a room for him to live in.[25] It is less clear why a bridge should have attracted recluses. The Pont-d'Estrouilhas (Haute-Loire) had its *domus recluze pontis* in 1359, and in 1398 there were two recluses du Pont-Neuf at Toulouse.[26] At Agen the town government decided on 7 March 1351 to obtain permission from the bishop and chapter to demolish the little house of the recluse in front of St. George's church and use the lumber to build a new cell (*domus recluse*) on one end of a pier across from the town weights located on the other end.[27] The word *recluse* is related to *reclusage*, meaning cloister, and since there were convents along the streets of medieval towns, there was no reason why there should not be a recluse living on a bridge pier. Perhaps the question of alms was involved, for a bridge was a favorite site for their collection. For example, the word "hermitage," like "recluse," implies solitude. Yet there was a hermitage at Périgueux in 1492 at the entrance to the Pont-de-la-Cité on the main road to Bordeaux. Since the favorable location procured the hermit a generous sum in alms, the municipality charged him 5 *l.* annually.[28]

Towers had been a feature of medieval bridges at least since the ninth century, and their military role was accentuated in the fourteenth century. The fact that the seal of the town of Lyons for 1320 showed towers whereas that of 1271 had not was a forecast of things to come (Fig. 15). The Hundred Years' War stimulated the fortification of towns and of bridges as part of the ramparts. Agen and Tours built towers to protect their bridges.[29] The func-

24. Ibid.
25. J. Artières, *Documents sur la ville de Millau* (Millau, 1930), p. 81, #180 (1342).
26. Pont-d'Estrouilhas (Haute-Loire), com. Aiguilhe and Espaly-Saint-Marcel: *Dic. top. Haute-Loire*, s.v. C. Douais, "Des Fortunes commerciales à Toulouse et de la topographie des églises et maisons religieuses de Toulouse d'après deux testaments," *Mém. Midi* 15 (1891), 35.
27. A. Magen, ed., *Jurades de la ville d'Agen* (Auch, 1894), p. 227.
28. F. Villepelet, "Le Moulin du pont de la Cité, en 1607," *Bull. Périgord* 47 (1917), 182, n. 5.
29. Vidal, *Douze Comptes*, 1:57, 98 (1368–1369). J.M.A. Delaville-Le-Roulx, ed., *Registres de comptes municipaux de la ville de Tours* (Tours, 1878–1881), 1:136, 366; 2:220, 224, 270, 271, 277–279.

tion of bridge towers may be illustrated by accounts of fighting at St. Cloud and at Orleans. During the Armagnac-Burgundian struggle there was a spirited engagement between the two factions involving the bridge of St. Cloud, which was defended by a tower. At the beginning of November 1411 the town of St. Cloud was in possession of the Armagnacs, who on November 10 were driven out of town by the Burgundians to the bridge tower. When an excessive number of Armagnacs rushed onto the drawbridge, it collapsed, and many were drowned in the Seine. The tower, nevertheless, was not taken, as it had been reinforced.[30]

The fortifications of the bridge at Orleans were elaborate (Fig. 4). At the town end the Porte Sainte-Catherine was flanked by the Châtelet, castle of the dukes of Orleans, and at the Sologne end, supported by the first two piers of the bridge, was a fort, Les Tourelles. Just before the arrival of the English army the residents had built a fortified area surrounded by a ditch called the *boulevard du Portereau.* It was a standard defensive measure for a town awaiting an attack to destroy the suburbs.[31] When the English reached the bank of the Loire opposite Orleans, they found the suburb of Portereau still in flames. They attacked the *boulevard* and tunneled underneath it, assaulting it on 21 October 1428. Two days later the Orléanais set fire to the fortifications, and the next day they abandoned the Tourelles as untenable. The English nevertheless were unable to force a passage across the bridge, perhaps because they had lost their commander, the Earl of Salisbury, and because of a shortage of manpower. When Joan of Arc arrived in Orleans, she fired the people's enthusiasm to the point where the leaders organized an attack on the fortifications of the left bank. On 7 May 1429 the French took the *boulevard* and the fort of the Tourelles.[32]

The damage to the bridge was greatly exaggerated in requests to the royal government to grant permission to levy an *aid* or tax to pay their debts and make repairs. Letters of King Charles VII of 16 January 1430 (n. st.) state: "Our well-beloved bourgeois, residents, and inhabitants of the town of Orleans have represented to us that to maintain their loyalty to us and to resist the damnable enterprises of the English, our ancient enemies and adver-

30. T. Basin, *Histoire de Charles VII*, ed. C. Samaran (Paris, 1933), 1:30–31. *La Chronique d'Enguerran de Monstrelet*, ed. L. Douët-D'Arcq (Paris, 1857–1862), 2:203–206.
31. Compare the actions of the town council of Agen; Magen, *Jurades*, pp. 4, 33, 40, 68, 72, 105.
32. R. Boucher de Molandon, A. de Beaucorps, *L'Armée anglaise vaincu par Jeanne d'Arc sous le murs d'Orléans* (Orleans, 1892).

saries, who for eight months or thereabouts have besieged the town," they had gone into debt, their houses had been destroyed, and "the bridge of the said town with its towers has been in great part demolished."[33]

Impairment of the bridge does not seem to have been on the scale described in the above royal letters. Bombards and cannons had indeed shot holes through the roofs and walls of the buildings, and four arches had been cut, two north of the Tourelles, one next to the Belle Croix, and one adjacent to Saint-Antoine.[34] Apparently the English had been responsible for the destruction of two arches, and the Orléanais of another two as defensive measures. Nevertheless, on the departure of the English the Orléanais almost immediately restored the bridge to use by placing beams and planks between piers to substitute for the fallen arches.

The old bridge at Orleans crossed the Loire where there was a sandy island, and the distance between banks amounted to some 350 meters. At the beginning of the fifteenth century the bridge consisted of twenty arches. On the eighth pier, counting from the southern bank, was the Belle Croix, erected at the expense of the bridge and hospital in the years 1405–1407.[35] There were fourteen piers between the left bank and the island, the eastern end of which, the site of the chapel, was called the Motte Saint-Antoine. The island's western end, where both the hospital and a fish market were situated, was named the Motte des Chalans Percées or des Poissonières. Before the flood of 27 February 1435 (n. st.) there were six piers between the Motte Saint-Antoine and the right bank of the river, and it was in this part of the river that the mills were placed.

Repairs began by the *proviseurs* of the bridge hiring workmen to remove the rubble and then turning their attention to the pressing problem of restoring the buildings. Repair of the structures was essential for the financial well-being of the bridge, partly because of rents from the houses but also because donations were in part dependent on the functioning of the chapel and of the hospital of Saint-Antoine.[36] It was longer before the arches damaged during

33. Archives du Loiret, CC 967, fol. 2r ff.
34. A. Collin, *Le Pont des Tourelles à Orléans* (Orleans, 1866), p. 497.
35. *Inventaire-sommaire arch. com. Orléans*, 1:193.
36. Manuscript CC 960 of the communal archives of Orleans is chiefly taken up with details of sums paid by the *proviseurs du pont* in 1328–1431 to masons, carpenters, roofers, and laborers to repair the bridge properties. Parts of the structures were in stone, parts in wattle and daub construction. Before the impending siege many of the beds, the linens, and also a barrel of wine had been transferred from the hospital and bridge to the town, where rooms had been rented for them.

the siege were rebuilt in stone. The roadbed of the arch beyond the Belle Croix was paved in 1433–1434,[37] but in October 1436 the flooring of the arch of the Cross still consisted of planks.[38] The proctors of the town of Orleans appealed to the governor for aid in paying for the restoration of the bridge, but in 1447 the lieutenant of the duchy's governor decided that it was the town that must pay for repairs on the grounds that the destruction of the arches had been the fortune of war.[39]

The damage to the bridge during the English siege was not nearly so extensive as that due to the floods of 1435. The war had cut four arches of the bridge, but ice and the rampaging waters of the Loire swept away six arches and four piers between the Châtelet and the Motte Saint-Antoine, leaving only the arch containing the mills belonging to the duke of Orleans. The ruined stretch had included the master arch used for navigation and five houses built on the bridge. The fact that the flood damage was confined to the channel between the Motte Saint-Antoine and the Châtelet was probably due to a submersible dam across the rest of the river bed. Evidently the mills in the ducal arch protected it from the flood waters. After the inundation, fallen masonry blocked the navigational channel. On the grounds that they prevented the passage of boats, large or small, eleven piles were pulled out from beneath the one remaining arch between the Motte Saint-Antoine and the town, and the mills were temporarily removed.[40]

An improvised wooden bridge was thrown across the gap, and the citizens of Orleans appealed again to the king for power to collect extraordinary aids. The royal government granted an *aid* called *les nouvelletez* to be levied on merchandise by land and water. The accounts for the period 2 February 1435/6 to 2 February 1436/7 were specifically designated as "Compte de Michelet Filleul pour le fait du Rediffiement du pont."[41]

In accordance with the purpose of the *nouvelletez*, the administrators repaired parts of the bridge not damaged in the floods, as well as those damaged in the catastrophe. For example, the third arch beyond Saint-Antoine was rebuilt.[42] At this time an important change was made in the bridge, for the

37. P. Mantellier, "Mémoire sur la valeur des principales denrées et marchandises qui se vendaient ou se consommait en la ville d'Orléans, au cours des XIVe, XVe, XVIe, XVIIe, et XVIIIe siècles," *Mém. Orléanais* 5 (1862), 343.
38. Archives du Loiret, CC 967, fol. 17r. In October 1436 a carpenter, his brother, and his valet were paid for having replaced and nailed planks on the arch of the Cross.
39. Collin, *Le Pont*, pp. 465, 497.
40. Archives du Loiret, CC 967, fols. 14v, 18r. See also Collin, *Le Pont*, p. 496.
41. Archives du Loiret, CC 967.
42. Archives hospitalières du Loiret, Ser. A III A 1, fol. VI^{XX}Xv.

three arches between the arch of the ducal mills and the Châtelet were reduced to two. The acccunts note payments for having the piles surrounding the unwanted pier pulled from the river bed. This meant that where there had been seven arches between the Motte Saint-Antoine and the Châtelet, there were now only six.

Typical features of a medieval bridge are shown in pictures in a manuscript life of St. Denis presented to King Philip V in 1317 (Fig. 16).[43] In the center rises a great crenelated tower through which one passes as he crosses the bridge. Each gate is fitted with a portcullis so that passage can be prevented at will. On the right is shown one shop, on the left two, occupied by moneychangers and jewelers. Beneath the roadbed there are seven round masonry arches. In one and in only one of the bridge illustrations are mills shown (Fig. 17). In each of three arches the millstones are shown housed in a small building on piles, and access is by means of a ladder from a rowboat. The mill wheel in each case is undershot, and one end of the axle appears to be supported in the side of the arch.

Medieval illustrations of bridges are sufficiently rare so that historians have been inclined to read all they can into the few that exist. Various writers have assumed that the bridge in the life of Saint Denis depicts the Grand Pont or the Petit Pont or the Pont-aux-Meuniers, and Mrs. Egbert has advocated the ingenious idea that "the bridge" is not one bridge but two distinct bridges.[44] She assumes that the view is upstream and that the bridge on the left represents the Grand Pont, the bridge on the right the Petit Pont, while the tower in the middle represents a portion of the walls of Paris. For a variety of reasons this hypothesis will not bear examination. In the first place, the miniatures show us pointed cutwaters, and we know from seventeenth-century excavations of the remains of the Grand Pont, destroyed in 1296, that the cutwaters were pointed upstream and square downstream. Therefore

43. H. Martin, *Légende de Saint Denis. Reproduction des miniatures du manuscrit original presenté en 1317 au roi Philippe-le-Long* (Paris, 1908). V.W. Egbert, *On the Bridges of Mediaeval Paris. A Record of Early Fourteenth-Century Life* (Princeton, 1974). Interestingly enough, the two books illustrate a change in taste in the last sixty years. Martin in 1908 published the whole of each miniature, for he was intrigued by the life of St. Denis, but Mrs. Egbert in 1974 reproduced in most cases only the bottom third, for her interest centers in genre scenes.

44. Martin, *Légende*, p. 31. G. Fagniez, *Etudes sur l'industrie et la classe industrielle à Paris au XIIIe et au XIVe siècle* (Paris, 1877), p. 157, n. 1. R. Bennet, J. Elton, *A History of Corn Milling* (London, 1898–1904), 2:74. Le Roux de Lincy, Tisserand, eds., *Paris*, p. 54, give the title of the miniature with the mills as representing the Grand Pont, but in the index they refer to it as the Petit Pont. Egbert, *On the Bridges*, p. 24.

in the illustrations from the life of Saint Denis we are looking not upstream but downstream, so that the Petit Pont should be on the left. However, the identification with the Petit Pont cannot be correct, both because the left-hand bridge is too long for the Petit Pont (which evidence indicates had only three arches) and because a moneychanger's booth, confined by royal ordinance to the upstream side of the Grand Pont only, appears on that bridge. Then again, if the tower in the middle of the miniatures is really part of the fortifications of the Ile-de-la-Cité, this eliminates half the island. On the other hand, the bridge in the illuminations cannot be the Pont-aux-Meuniers, which was bordered only by mills, not by stores or workshops. It cannot be the Grand Pont in 1317, because all the Parisian bridges of the time were in timber, unlike the bridge in the manuscript. Both the Grand Pont and the Petit Pont had been washed out on 21 December 1296 by a flood which an eyewitness reported to have subsided fully only by 25 March 1297.[45] A wooden version of the Petit Pont was passable by 28 April 1297, and a Grand Pont in timber by 12 November 1297. In actuality, the bridges in the miniatures are not a faithful representation of any bridge at Paris in the early fourteenth century.

The new and wooden Grand Pont was built on a somewhat different site from that of the old stone structure. It began, as had its predecessor, at the Tour de l'Horloge at the royal palace, but instead of ending at the Châtelet on the right bank, it terminated somewhat upstream from the former bridge at the Rue de la Vieille Joallierie next to Saint-Leuffroy. It has been suggested that Philip the Fair displaced the bridge so as to transfer it from the jurisdiction of the cathedral chapter to his own.[46] The houses and workshops were rebuilt on the new bridge, and Philip IV in 1304 re-established the moneychangers on the Grand Pont between Saint-Leuffroy and the great arch.[47] Across from the row of moneychangers were the jewelers, and on the three arches next to the royal palace there were drapers.

It was a different matter with the mills formerly downstream from the Grand Pont and destroyed in the flood. They were rebuilt on their former sites. At the Grand Pont in the thirteenth century there was a mill in ten of

45. H. Sauval, *Histoire et recherches des antiquités de la ville de Paris* (Paris, 1724), 1:225. A. Vernet, "L'Inondation de 1296–1297 à Paris," *Mém. Féd. Paris* 1 (1949), 47–56.

46. Borrelli de Serres, "L'Agrandissement du Palais de la Cité sous Philippe le Bel," *Mém. Paris* 38 (1911), 82. A. Berty, "Recherches sur l'origine du Grand Pont de Paris," *Rev. arch.* 12 (1855), 19.

47. Borrelli de Serres, "L'Agrandissement," p. 66.

the eleven arches to the north of the great arch reserved for navigation and in three arches to the south of it, each of the mills being downstream from the bridge.[48] According to Etienne Boileau in his *Règlemens sur les arts et métiers* at the end of the thirteenth century, there were enough millers of the Grand Pont to form their own corporation and be listed among the guilds of the city of Paris.[49] After the catastrophe of 1296 mill owners rebuilt their mills where they had been, and each constructed that part of the bridge necessary to his own mill. A pronouncement by the provost of Paris fixed responsibility for the upkeep and listed mill owners of 1323. The first arch from the right bank next to Saint-Leuffroy was the responsibility of all proprietors of mills, inasmuch as they used it to transport their grain and flour. The second housed the mill of Guillaume le Meunier, the third and fourth the mills of the chapter of Notre Dame, the fifth that of St.-Ladre, the sixth that of Saint-Germain-l'Auxerrois, the seventh that of the Temple, the eighth that of Saint-Martin-des-Champs, the ninth that of Saint-Magloire, the tenth that of Saint-Merri, and the eleventh that of Sainte-Opportune. The twelfth arch was the great arch open for navigation.[50] Here the Pont-aux-Meuniers ended without crossing the Seine, for its purpose was to serve mills rather than to afford a passage across the river.

The mills "under" the old Grand Pont were even farther downstream from the new. Bridge owners preferred a considerable distance between bridges and mills, as a mill wheel too near a pier caused scouring. At Agen in 1189 and at Saint-Maur-des-Fossés in 1205 no mill was to be built too near a bridge for fear of damage.[51] At Orleans a floating mill causing the bridge pier to deteriorate was removed in 1405.[52] Here again the illustration in the life of St. Denis is probably incorrect in representing the mill-wheel axle as fixed among the stones of the pier (Fig. 17). It should have been downstream supported by the piles of the millrace.

Property rights probably played a large part in the decision to rebuild the mills on their former sites. For example, the mills of Saint-Merri and Sainte-Opportune had been ordered destroyed by the king to relieve pressure on the

48. Ibid., p. 64. J. Guérout, "Le Palais de la Cité à Paris," *Mém. Féd. Paris* 1 (1949), 94, n. 2.
49. E. Boileau, *Les Métiers et Corporations de la ville de Paris, XIIIe siècle*, ed. R. de Lespinasse and F. Bonnardot (Paris, 1879), p. 16.
50. Fagniez, *Etudes*, p. 160, n. 2.
51. G. Tholin, "Les Ponts sur la Garonne," *Rev. Agenais* 5 (1878), 440. G. Bousquié, "Histoire du pont de Saint-Maur," *Mém. Féd. Paris* 4 (1952), 331–332.
52. Archives du Loiret, CC 920, fol. 13r. Collin, *Le Pont*, p. 434.

bridge during the flood of December 1280, in which two arches of the Grand Pont were washed out. Philip IV hesitated to allow rebuilding of the two mills, presumably because of the threat to the bridge, but he was unable to resist ecclesiastical sanctions.[53] The mills were rebuilt and no doubt played their part in bringing on the destruction of the Grand Pont in 1296. To have moved the mills to positions below the new timber Grand Pont would have involved the purchase of new sites and might not have been practical. Convents owning the river bottom at Paris tenaciously clung to their rights to every foot of water. Saint-Magloire charged a sum for the driving of every pile in the Seine between the Ile-Sainte-Marie and the old Grand Pont.[54] There were quarrels between mill owners whose races were separated by rows of piles as to whether one had encroached on the other's water. For example, in 1325 sworn masons and carpenters of the city of Paris went to the Pont-aux-Meuniers to settle a dispute between the convents of St.-Magloire and Saint-Martin-des-Champs.[55]

There were a number of reasons why millers preferred to locate mills near bridges. Cutwaters and piers of a masonry bridge could be incorporated into a mill race; also, the great stone piers, by restricting the freeway, tended to accelerate the current and at some places could be incorporated into a dam. Sometimes, but rather rarely, the mill could be built on the bridge with convenient access from the roadbed, as shown in the illustration of Corbeil mills and bridge (Fig. 18). At Paris after 1296 the owners, by rebuilding their mills on the old sites, were able to make use of the remains of the piers of the old stone bridge.[56] It is improbable that the mills could have been reached from the old Grand Pont before 1296,[57] because space on the bridge was at a premium. Besides, it was usual, and evidently more profitable, both at Paris and elsewhere, for the owner of a house "on" a bridge arch and a mill "in" the same arch to rent the two to different persons. This the Templars did at the Grand Pont in the thirteenth century.[58] The description of a mill as "in" an

53. Borrelli de Serres, "L'Agrandissement," p. 82.
54. A. Terroine, L. Fossier, eds., *Chartes et documents de Saint-Magloire* (Paris, 1966), passim, esp. 2:388.
55. Ibid., 2:448. In 1407 a row of piles belonging to St. Magloire encroached upon the territory of St. Merri, and an accommodation was reached whereby both convents used the same *jouée* (walkway on a row of piles): Fagniez, *Etudes*, p. 159, n. 6.
56. Berty, "Recherches," p. 210.
57. Borrelli de Serres, "L'Agrandissement," p. 80, states that millers of the Grand Pont could no longer make use of ladders to reach their mills. He gives no evidence for this statement.
58. B.E. Guérard, ed., *Cartulaire de l'église Notre-Dame de Paris* (Paris, 1850), 4:395 (for 1248) and 4:527 (for 1240). See also A. Champollion-Figeac, *Droits et usages*

arch did not mean that it was under the roadbed of the bridge. In 1212 there was a mill in an arch of the Petit Pont which was more than 10 *toises* (19.49 m.) upstream from the bridge.[59] It is not surprising to read that one at least of the mills at the Petit Pont had its own arch so that it could be reached conveniently from the bank.[60]

The Pont-aux-Meuniers, far from being unique, was only one of a number of mill bridges at Paris and elsewhere in France. It is famous because it served more mills than others and because in the sixteenth century the bridge extended across the navigable arch and in front of the mills to the Ile-de-la-Cité.[61] A view of the city of Paris in that period seems to show the Pont-aux-Meuniers as a bridge like other bridges, obscuring the fact that it continued to be for the exclusive use of millers and their patrons.

Access to mills could be furnished by boat, as shown in the picture in the life of St. Denis (Fig. 17), but it seems to have been usual for Parisian mills to be served by special walkways constructed of piles with planks on top. The sides of the mill race ordinarily could be used as walkways so that one could approach the mill race from either side.[62] This was especially important when the mill race served also as a fish pond (see Fig. 18, Corbeil). Mills at Paris are described as possessing bridges, arches, or *itinera.* For example, in 1357 the brothers of the hospital of St. John of Jerusalem had a "certain pont ou alee" to reach their mill at the end of the Grand Pont at the side of the palace. Their bridge was adequate to accommodate horse and pedestrian traffic.[63] Above the Grand Pont and unconnected with it were several mill bridges or arches. A Saint-Magloire charter of 1323 mentions: "In the water of the Seine, at Paris, a mill with an arch located above the mill bridge of the bishop of Paris at the place they call Planches de Mibray. . . ."[64] In 1317 the episcopal *pont-des-moulins* was allowed to be eight feet wide and forty-five to fifty piles long.[65]

---

concernant les travaux de construction (Paris, 1860), p. 153, for 1270. E. Lemaire, ed., *Archives anciennes de la ville de Saint-Quentin* (Saint-Quentin, 1888), 1:31. In 1237 the house above the mill was occupied by a blacksmith.

59. Guérard, *Cartulaire*, 4,1:141.
60. L. Halphen, *Paris sous les premiers Capétiens* (Paris, 1909), p. 3, n. 2.
61. G. Braun, W. Hogenberg, *Civitates orbis terrarum, 1572–1616* (Amsterdam, 1965).
62. Guérard, *Cartulaire*, 4,3:224. In 1294 the bishop of Paris was adjudged to possess the right to use a pathway on top of the piles between his bridge and the Moulin Robin de Corbeil, on the Grand Pont (actually just downstream), on the condition that he pay half of the maintenance costs.
63. Fagniez, *Etudes*, p. 18, n. 2.
64. Terroine and Fossier, *Chartes*, 2:407.
65. Ibid., 2:324 ff.

The bishop's bridge mill at Planches de Mibray may have been responsible for the tradition, repeated by many writers of the modern period,[66] that there was a bridge of the Planches de Mibray preceding the Pont-Notre-Dame. Mibray was a place name appearing in twelfth-century documents, and the Parisian taille of 1292 lists a "ruele des Planches de Mibrai" on the Right Bank leading to the Seine.[67] It is noteworthy that the earliest writers noticing a bridge of the Planches de Mibray do so in the past tense. Raoul de Presles, writing in 1370–1375, and René Macé, a contemporary, mention that there had been a timber bridge at that point.[68] Nothing can be deduced from the journal of the bourgeois of Paris as to a bridge at the Planches de Mibray before 1412, when the town of Paris secured permission from the convent of Saint-Magloire to build a bridge.[69] The bourgeois of Paris merely uses the term "Planches de Mibray" as a place name. It seems certain that the Pont Notre Dame was the first bridge at that site for the use of persons wishing to cross the Seine.

One can today visit the Pont Valentré at Cahors, if one wishes to see fortifications, or the Pont-Saint-Bénézet at Avignon, if one prefers the view of a deserted chapel on a medieval bridge. However, it is the medieval bridge at Orleans, now a ghost bridge, traces of whose venerable piers may still be seen in the bed of the Loire during times of drought, that epitomizes the history of bridges in the Middle Ages. It was built of masonry in the great age of stone bridges–the twelfth century–and endowed by charity as a pious work. It underwent siege, fire, and flood, but the will to maintain it continued into the eighteenth century, when higher levees along the Loire rendered its low level impractical. Its appearance emphasized the functions of a medieval bridge. It had its own chapel, hospital, fortifications, houses, shops, workshops, mills, and markets. In addition to its part in town life and prosperity, the bridge with its great length and impressive towers must have been a commanding presence. The Orléanais were inordinately proud of their bridge. They boasted of "la grant renomee qui est du pont dans l'universel monde."[70]

66. J. de Saint-Victor, *Tableau historique de Paris* (Paris, 1808–1811), 1:173–174. C. Duplomb, *Histoire génerale des ponts de Paris* (Paris, 1911–1913), 1:201–202. Le sieur Jaillot, *Recherches critiques, historiques et topographiques sur la ville de Paris*, new ed. (Paris, 1782), 1:191. J.–A. Dulaure, *Histoire de Paris* (Paris, 1852), p. 188. Dulaure notes the mill bridge at Planches-de-Mibray.
67. Lasteyrie du Saillant, ed., *Cartulaire général de Paris* (Paris, 1887), 1:213 (1122), 1:239 (1129). P.H.J.F. Géraud, *Paris sous Philippe le Bel* (Paris, 1837), p. 301.
68. Géraud, *Paris*, p. 387. Le Roux de Lincy, Tisserand, *Paris*, p. 109.
69. A. Tuetey, ed., *Journal d'un bourgeois de Paris* (Paris, 1881), pp. 30, 31.
70. Collin, *Le Pont*, p. 631, letters patent of Charles VII, 24 December 1435 and 14 May 1436.

IV

THE BUILDERS AND THEIR BRIDGES

10

Bridge Construction and Bridgebuilders

MEDIEVAL FRENCH BRIDGE CONSTRUCTION was strongly influenced by the particularism of the era. Most materials were gathered in the vicinity, and most workmen were local men. If a mason came from outside the community, it was usually from some town not too far away or perhaps from the next province. Neither do ideas seem to have spread rapidly or over great distances. The medieval French bridges which show the most pronounced Roman influences were located in the South, where builders could readily see the ancient models.

The typical medieval dichotomy between theory and practice, between learned treatises on the one hand and artisans' secrets on the other, is illustrated only to a small extent in the history of French bridges. One reason was a singular lack of medieval literary works devoted to bridge construction. The interest of writers seems to have been limited to passing references to materials used in bridge construction or to a discussion of machines which could be employed for that purpose. For example, in the thirteenth century Villard de Honnecourt's sketchbook includes a drawing of a machine for sawing off the tops of piles.[1] Even when the information was available, it was not necessarily used. An eighth-century manuscript of the *Mappae Clavicula*[2] includes a formulary, based on ancient practices, for laying in water the foundations of buildings, a method applicable to founding bridge piers. This method, although available in a Carolingian manuscript, seems not to have been employed in building the twelfth-century bridges across the Loire. Probably in the thirteenth century and certainly in the fourteenth and fifteenth centuries the method of founding piers on piles was in use.

1. *The Sketchbook of Villard de Honnecourt*, ed. T. Bowie, (Bloomington, 1959), p. 130, Pl. 60 (CXLV).
2. V. Mortet, "Un Formulaire du VIIIe siècle pour les fondations d'édifices et de ponts d'après des sources d'origine antique," *Bulletin monumentale* 71 (1907), 442–465.

In the medieval period by far the most important material in bridge construction was wood. This was true even in the case of stone bridges, where scaffolding and centerings required large amounts of timber. The enormous quantities of wood needed came as a shock to those responsible for bridge construction. At Albi eighteen cartloads of timber were insufficient to provide the lumber used in replacing one masonry cutwater of brick and stone, and at Romans thirty-three fir trees were necessary to reconstruct in stone a single arch.[3] Perhaps it is not surprising that in 1334 the abbot of Chassagne, anxious to rid his convent of the burden of trying to rebuild the Pont-de-la-Guillotière at Lyons in stone, complained that the convent's forests had been devastated by the demands of the bridge as by a war.[4] The necessity of a regular supply of timber for repairs was so well recognized that a number of bridges had the right to cut trees for this purpose in a specific forest, as for example, Montereau in 1037 and Gournay-sur-Marne in 1190.[5]

There was a direct continuity between medieval and eighteenth-century construction practices and use of materials in bridgebuilding, as shown in the manual of Henri Gautier, *Traité des Ponts.* For example, both periods were in accord on the kinds of wood to be used. Gautier advised against the use of firs, except underwater where fir fails to rot. It cannot support a great weight nor does it last long exposed to air.[6] At Romans firs and at Albi poplars and nut trees were used only for centerings, scaffoldings, and temporary bridges during reconstruction of stone arches.[7] The ideal wood for bridges, according to Gautier, was chestnut, which will carry great weight and does not rot in the air.[8] At Agen, where much of the bridge was planned to be permanently of timber, a contract of 1381 specified that the wood was to be "good and sufficient chestnut wood."[9] Chestnut was also used at Albi, and at Lyons in 1454 piling under a pier foundation was required to be of chestnut.[10] Alder,

3. Archives communales d'Albi, CC 167, fols. I verso–xiii verso. U. Chevalier, "Notice historique sur le pont de Romans," *Bull. Drôme* 2 (1867), 316.
4. M.–C. Guigue, *Cartulaire municipal de la ville de Lyon* (Lyons, 1876), p. 169 ff.
5. R. Poupardin, ed., *Recueil des chartes de l'abbaye de Saint-Germain-des-Près* (Paris, 1909), 1:214. A. Catel, M. Lecomte, eds., *Chartes et documents de l'abbaye cistercienne de Preuilly* (Paris, 1927), p. 81, in 1190–1191. J. Depoin, ed., *Recueil de chartes et documents de Saint-Martin-des-Champs* (Paris, 1912–1921), 3:#515. See also E. Boutaric, ed., *Actes du Parlement de Paris*, 1 (Paris, 1863), 442–443.
6. H. Gautier, *Traité des ponts*, 4th ed. (Paris, 1765), 1:55.
7. Archives communales d'Albi, CC 168, fols. xliii verso, lxxiiii recto.
8. Gautier, *Traité*, 1:55–56.
9. O. Fallières, "Le Pont d'Agen en 1381," *Congrès arch.*, 68e session, (Agen, 1901), p. 437.
10. M.–C. Guigue, "Notre-Dame de Lyon," *Mém. Lyon* (1874–1875), p. 252.

a favorite wood for London Bridge before the stone structure was built, was specified in 1347 for the planking of the bridge at Agen.[11]

On the other hand, my researches have not uncovered evidence that medieval people adhered to the dictum, expressed by Gautier, that trees should be cut only after shedding their leaves and before the buds appeared.[12] At La Robertière (Eure) when trees were felled and a bridge built in Lent 1346, it was evidently the impending arrival of a dignitary which forced work earlier in the year than was usual.[13] Trees seem to have been cut whenever convenient, before or during the bridgebuilding season. There was a well-defined period of the year when bridge construction was carried on. Gautier merely stated that operations should be terminated before the autumn rains and resulting floods, and for the same reasons a ninth-century capitulary ordered work completed before Saint Andrew's day (November 30).[14] In my researches work seems rarely to have begun before the middle of June, and at Orleans the pile-driving season started in late June and extended into September or October. August, because of low water, was declared the best season at Pont-Saint-Esprit, and August and September were favorite months at Albi.[15]

Medieval people were thoroughly in agreement with Gautier that a stone bridge was much to be preferred to a timber one. Gautier believed that a masonry bridge should be built even if ten times as expensive as a wooden structure, because the difference in maintenance costs would justify the initial outlay.[16] A timber bridge was always in need of repairs. Medieval people, however, seem to have found that stone bridges, too, required continuous maintenance, partly because of a method of founding piers then in use. Ordinarily local stone was used. For example, builders of the Pont-Saint-Esprit procured stone from a quarry near the eastern end of the bridge. At Romans and at Nyons the material was indigenous tufa, at the Grand Pont of Paris calcareous limestone.[17] A mixture of stone and brick was used in

11. A. Magen, ed., *Jurades de la ville d'Agen* (Auch, 1894), pp. 125–126.
12. Gautier, *Traité*, 1:53.
13. The manor of La Robertière was near Saint-Georges-Sur-Eure (Eure), arr. Evreux, cant. Nonacourt: L. Delisle, *Actes Normandes de la Chambre des Comptes sous Philippe de Valois, 1328–1350* (Rouen, 1871), p. 338.
14. Gautier, *Traité*, 1:32. A capitulary of Louis the Pious in 825 ordered bridge repairs completed before Saint Andrew's day (November 30) for fear of floods: MGH Capit., 1:307.
15. L. Bruguier-Roure, ed., *Cartulaire de l'oeuvre des église, maison, pont et hôpitaux du Saint-Esprit*, *Mém. Acad. Nîmes*, ser. 7, vol. 12 (1889), Annexe, p. 6.
16. Gautier, *Traité*, 1:57.
17. Chevalier, "Notice," pp. 316–317. P. Séjourné, *Grand Voûtes* (Bourges, 1913–

bridges at Albi, Agen, and Toulouse. At Albi it was a shortage of quarries that caused the use of bricks to supplement stone. The same was true at Toulouse, where a pier built after 1480 was of brick, and its cutwater had a facing on the lower and middle part of limestone imported from the Pyrenees.[18] To transport stone such a distance was unusual, but evidently the superior resistance of the limestone to water made it worthwhile.

Initiation of repairs was frequently the responsibility of the day to day administration of the bridge, which was sometimes in charge of a commission, as for example, the *proviseurs* at Orleans, sometimes of an *operarius*, as at Agen. The word *operarius* (in Provençal *obrer de pont*) was loosely used to mean the man in charge of financial affairs or of maintenance, or the contractor making repairs. It was common to sign a contract for construction. In the thirteenth century a contract was let at auction to rebuild a small wooden bridge at Oulchy-le-Château (Aisne).[19] There are a number of examples of contracts from the fourteenth and fifteenth centuries—from Agen, Albi, Lyons, Orleans, and Toulouse, among others. At Orleans and Lyons efforts were made to assure publicity by employing the town crier. For example, at Lyons he was ordered to announce that anyone wishing to undertake to rebuild two fallen arches of the Pont-de-la-Guillotière should appear at Saint-Jaquême on 10 January 1415, when the contract would be let.[20] At Orleans, too, care was taken in 1436 that specifications for the new arch adjacent to the Châtalet and the letting of the contract should obtain an audience. Twelve copies of the plans were made and displayed in the public squares "so that everyone might see them."[21] A sergeant cried the letting of the contract for the arch to take place on 7 May 1436.[22] Awarding a contract for bridge construction was important enough to require the presence of numerous citizens. The Parlement of Toulouse insisted that the aldermen (*capitouls*) of the town assemble the great council to approve a contract to reconstruct the Pont-de-la-Daurade in 1477.[23] Similarly, at Nyons (Drôme) in 1361 a contract

1916), 2:10. H. Sauval, *Histoire et recherches des antiquités de la ville de Paris* (Paris, 1724), 1:225.

18. G. Astre, "Techniques médiévales et modernes: les matériaux du pont médiéval de la Daurade," *Annales du Midi* 63 (1951), 351–352, "Beau calcaire nankin ou pierre jaune de Roquefort, celle qui provient du terrain maëstrichtian des Petites Pyrénées."
19. Oulchy-le-Château (Aisne), arr. Soissons. A. Longnon, *Documents relatifs au comté de Champagne et de Brie, 1172–1361* (Paris, 1901–1914), 3:66.
20. Guigue, "Notre-Dame de Lyon," pp. 249–250.
21. Archives du Loiret, CC 967, fol. 25r. See also fol 17v.
22. Ibid., fol. 23r.
23. Archives dépt. Haute-Garonne, B 4, fol. 355. See also B 1, fol. 226; B 2, fols. 163, 169; B 3, fol. 21; B 4, fols. 264, 309, 369; B 6, fol. 21.

to construct an arch was signed by a mason and 153 inhabitants assembled in the town council.[24]

At Agen in the years 1345–1355 the problem of securing competent supervision for bridge construction was a very difficult one for the town government. The *operarius* in charge of the bridge over the Garonne was ordinarily a consul or bourgeois of the town, frequently at the same time a clerk or notary. He received and disbursed bridge funds, paying the contractors and overseeing the work. Almost every year there was a new *operarius* or sometimes *operarii.* Ordinarily they received a salary, but in 1352 at a time of acute financial embarrassment for the town a consul and two masters, one a notary, volunteered to supervise bridge affairs without payment except for traveling expenses.[25] Evidently the informal method of volunteers managing the bridge was unsatisfactory, and on 24 January 1353 Master P. Bosiget, a notary, was appointed master of the *opus pontis* at a salary of 20 *l. t.* annually on the grounds that legacies were being lost, as there was no one to receive them.[26]

The troubles of Agen with its contractors were worse than those with the *operarii*, for if the latter were sometimes unconscientious, the former were in some cases inept or dishonest. The middle section of the bridge was carried away by a flood on 16 March 1347 and on July 7 the town contracted with Master Philippe de l'Arche to make a bridge of adequate height with an alder flooring, beginning at the entrance to the old bridge and extending to the bridgehead still standing on the opposite bank.[27] The actual work was to be done by Jean le Buder. Construction was to be completed within a year from St. Michael's day. The time limit was impractical, for later ten years were required to complete a "permanent" bridge of timber and stone. Perhaps the town council of Agen was not experienced in bridge construction, and certainly the entrepreneurs were not taking the contract seriously. The next month, on the death of Master Philippe de l'Arche, the town agreed on August 27 to allow his associates to construct the bridge.[28] This was a mistake, for they absconded to Toulouse, owing the town 500 *l.* On 17 April 1349 the town council decided to send to Toulouse to recover the money or the services of the contractors, but in vain, for nearly two years later the municipality was threatening court proceedings against Master Jean d'Estradoux, carpenter

24. Séjourné, *Grandes Voûtes*, 2:28 ff.
25. Magen, *Jurades*, p. 268.
26. Ibid., p. 302.
27. Ibid., pp. 125–126.
28. Ibid., pp. 129–130.

of Toulouse, one of the contractors.[29] Whether the *operarii* or contractors or both were ineffective, not much progress seems to have been made on the bridge until in January 1353 Master Pierre Bosiget was appointed *operarius.* He was able to secure a contractor who quickly built a service bridge to replace for pedestrians the bridge of boats; but only in 1381 did Agen obtain a contractor who actually completed the bridge within the decade.

In the fourteenth century contracts to build or repair bridges were made with a single man at Agen, Romans, Nyons, and Mâcon.[30] Ordinarily this man was either a mason, as for example at Céret (Pyrénées-Orientales) (ca. 1321–1339) and at Mâcon (1362–1370), or a carpenter, as at Agen in the fourteenth century or Ponts-de-Cé in the fifteenth,[31] where the bridges were predominantly of wood. It is true that the spheres of work of masons and carpenters overlapped, for the former helped put up scaffolds at Albi in 1408–1410, and the latter took pick or iron bar in hand to aid in demolishing the old arches. Nevertheless, it was exceptional that Guillaume de Pays described himself at Nyons in 1398 as both a mason and a carpenter.[32] It was much more typical of the medieval period to make a clear distinction between the two categories of artisans and to call in both for advice on bridge construction. At Pont-à-Mousson (Meurthe-et-Moselle) in 1230 the duke of Lorraine and the count of Bar were each required to send his own carpenter and his own mason to make an estimate of the cost of rebuilding a bridge which the nobles had destroyed.[33] In the fifteenth century two contracts were let at Albi and at Orleans for rebuilding a single arch, one to a carpenter to make the scaffolding and centering, one to a mason for the stone and brick work. At Albi there seems to have been some feeling between the masons and carpenters, for when the town was feeding the workmen during reconstruction of a cutwater in 1408, the masons insisted on a separate table, "because they did not wish to be with the above-named," the carpenters and their helpers.[34]

29. Ibid., pp. 161, 183, 220.
30. [L.] Michon, "Documents relatifs à l'histoire de la ville de Mâcon (1362–1367)," *Rev. Soc. Savantes*, 5th ser., 1 (1870), p. 182, n. 1. Séjourné, *Grandes Voûtes*, 2:26, on Nyons.
31. Séjourné, *Grandes Voûtes*, 2:21, n S_2. Michon, "Documents," p. 182, n. 1. A. Lecoy de la Marche, ed., *Extraits des comptes et mémoriaux du roi René* (Paris, 1873), p. 151.
32. Séjourné, *Grandes Voûtes*, 2:29.
33. *Layettes*, 2:191b. At Lyons in 1418 it was two masons and a carpenter that the consuls paid for rebuilding an arch: M.–C. Guigue, *Registres consulaires de la ville de Lyon* (Lyons, 1867), 1:97, 13 January 1418 (n. st.).
34. Archives communales d'Albi, CC 167, fol. x recto.

In general, most medieval French bridgebuilders were natives of the region where the bridge was built. Thus at Agen, so far as is known, all the *operarii* were residents of the town except for Mestre Jean d'Estradoux, carpenter of Toulouse. Such parochialism was not everywhere considered desirable, and in the fifteenth century the *proviseurs* of the bridge at Orleans arranged to have the letting of a contract to rebuild an arch publicized by the town crier at Sully and elsewhere.[35] At Nyons in 1398 the town employed a native of Romans, Guillaume de Pays, to build a stone bridge.[36]

In a number of cases attempts to secure advice from experts of other regions were made. At Romans the commissioners for the reconstruction of an arch sent the master of the works to Lyons in April 1388 to study plans for rebuilding the Pont-de-la-Guillotière in stone. Nor was this all. The Romanais also imported the master of the works of the bridge across the Rhone at Lyons to inspect the repairs to the bridge on the Isère.[37] Also, Albi in 1368–1369 had invited masons from the region to inspect the ruins of the fallen bridge tower and to give the commune advice on reconstruction.[38]

At Orleans the *proviseurs* of the bridge and hospital and members of the town government attempted to obtain the best advice and workmanship and to diffuse responsibility for decisions respecting the rebuilding of the bridge following the floods of 1435. The town government sent in vain to Tours to request the *maistre des oeuvres* of Touraine to come and give his "good advice and counsel" on the plans for the Orleans bridge.[39] Also, the *proviseurs* for the reconstruction of the bridge sent two quarts of wine to a merchant of Fougères (Ile-et-Vilaine) "qui avoit cognoissance en fait de maçonnerie," as recompense because he had given his opinion on the arch then building.[40] The *maistre des oeuvres de maçonnerie* of the duchy of Orleans was paid for promising to see to it that no stone or lime was used which was not profitable for the work and for having inspected the bridge on numerous occasions.[41] At Orleans advice was sought not only from outsiders but from residents. There, when work on the new arch between the duke's mills and the town was well under way, many carpenters and masons as well as sixty to eighty bourgeois and counsellors were summoned by sergeant to

35. A. Collin, *Le Pont des Tourelles à Orléans* (Orleans, 1866), p. 495, n. 3.
36. Séjourné, *Grandes Voûtes*, 2:29–30.
37. Chevalier, "Notice," pp. 315–316.
38. A. Vidal, ed., *Douze Comptes consulaires d'Albi du XIVe siècle*, 1 (Albi, 1906), 59.
39. Archives du Loiret, CC 967, fol. 14v.
40. Ibid., fol. 16v.
41. Ibid., fols. 25r, 26v.

appear on the bridge on 26 July 1436 to give their judgment as to whether the foundation for the pier was sufficiently deep. The decision was that it was not.[42] On August 3, carpenters, masons, notables, the provost and proctors of the town, and the lieutenant of the governor of Orleans were all assembled on the bridge, and this time their approval was forthcoming. They declared that the foundations were such that one could safely build on them. The workmen then drove piles shod with iron at the spot where the pier was to rise.

Inspections of bridges at Ponts-de-Cé, at Albi, and at Orleans were made by teams consisting not only of carpenters and/or masons but of government officials as well. At the Grand Pont the "jurés de la ville de Paris, maçons et charpentiers" gave expert advice on construction. For example, on 10 January 1326 the *jurés* were called in to decide the fate of a ruinous house on the Grand Pont. They declared that the house of Ysabeau de Tremblay was uninhabitable and in imminent danger of falling and dragging down the adjacent house and the king's bridge. Therefore the masons and carpenters recommended that it be torn down to the piles.[43]

Contracts varied as to whether the contractor or the town was to furnish the materials and as to which party should be responsible for paying the day laborers needed for the work. The remuneration of the contractor might be both in currency and in kind. For example, Master Thomas Hubert, mason, having contracted to rebuild an arch at Albi, was paid 140 *l.* plus amounts of wheat, wine, and cloth.[44] To an outsider it must have been a convenience to receive part of his sustenance. Sometimes the town also provided a bed, as at Albi in 1408–1410. The consuls of Albi preferred to pay the mason contracting to build an arch a lump sum, from which he was to hire the workmen. However, when it developed that Master Anthony Mauri was unscrupulous, the town found itself obliged to pay the wages of the workmen, who otherwise refused the job, "doubting that they would be paid by the said Master Anthony."[45]

Often the town provided the materials, as at Romans for rebuilding an arch and at Agen in 1347.[46] At Mâcon in 1367 in the case of repairs to an arch of the bridge the consuls agreed to provide the stone, lime, sand, two boats for working on the water, and a great rope. The mason, for his part,

42. Ibid., fol. 25v.
43. A. Terroine, L. Fossier, eds., *Chartes et documents de l'abbaye de Saint-Magloire* (Paris, 1966), 2:451–452.
44. Archives communales d'Albi, CC 168, fol. lxxv recto, in 1410.
45. Ibid., fol. lix verso.
46. Chevalier, "Notice," p. 317. Magen, *Jurades*, pp. 125–126.

promised to furnish the scaffolding and to pave the roadway of the arch.[47] The town was especially likely to furnish the timber, as at Agen, although frequently the contractor, if a carpenter, felled the trees, as at Albi. Usually the community was responsible for transportation, because residents were in a position to furnish the requisite manpower, oxen, horses, and carts. At Nyons the community in 1398 brought stones, sand, lime, and wood to the site.[48] At Donchéry (Ardennes) in 1322 when it was a question of rebuilding a flood-damaged bridge, it was agreed that the prior of the local convent was to send a boat to gather up as many of the beams carried downstream as possible and also to have new trees felled, that the community was to transport them to the site, the prior to have them sawed into planks, and the community to pay one-third the cost.[49]

When the workmen and materials had been assembled, it was time to begin construction. Details on rebuilding a wooden bridge are available for La Robertière on the Eure, a manor near Saint-Georges-sur-Eure. During Lent 1346 carpenters cut down and planed trees in the forest, and three-horse carts carried the timber to the site, where the piles were sharpened at the end and driven into the river bed. Afterwards tie beams were placed on them and wattles and wood laid on the crossbeams.[50]

In the case of a masonry bridge the first decision to be made concerned the method of founding the piers. In the twelfth century a number of bridges, including Old London Bridge as well as bridges as Blois, Tours, and Orleans, were built on starlings (*orgeaux*). To found piles on starlings made pile driving an essential and very large item in the expense of maintenance. Because the piers were placed on starlings surrounded by piles, new timbers had to be supplied yearly, not indeed at every pier, but here and there as needed. Thus at Orleans the *proviseurs* of the bridge and hospital purchased 400 piles in 1387, 175 in 1388, and 419 in 1389.[51] The timbers were sharpened but not shod with iron before being driven. Most were 2.92 to 3.89 meters long, with a few 4.54; and when the piles were 7.76 or 9.7 meters long, the accounts state that the workmen could not manage such long piles.

The difficulties with the long piles seem to be related to the size of the

47. Michon, "Documents," p. 182, n. 1.
48. Séjourné, *Grandes Voûtes*, 2:31–32.
49. G. Saige, J. Lacaille, eds., *Trésor des chartes du comté de Rethel* (Monaco, 1902–1916), pp. 633–634.
50. Delisle, *Actes Normandes*, pp. 338, 339.
51. Most of these piles were used for the starlings, but a few were driven along the banks of the Motte Saint-Antoine and along the edges of the river near the bridge, near the town, and near the *Portereau*: Archives du Loiret, CC 920, fols. 27r, 31r-v.

pile drivers. There were two types at Orleans in 1387–1389, the *hye* and the *engin.* The *hye*, which was little used at Orleans, had a windlass and was operated by six men. The *hye* was employed at the Pont Notre Dame at Paris, a timber bridge, for the ceremonial equivalent of laying the first stone of a masonry bridge. On 31 May 1413 King Charles VI struck the first pile of the new bridge with the *hye*, and afterwards his oldest son, the duke of Guienne, the dukes of Berry and Burgundy, and the lord of La Trémoille did likewise.[52] At Orleans the engine, which was employed steadily all summer, had no windlass and was powered by fourteen to sixteen men who "pulled on the ram."[53] Without the windlass, merely by letting go of the rope, the full force of the ram could be delivered suddenly and with maximum efficiency on the pile to be driven. The ram was made from an elm trunk, one such large trunk providing two rams. The ram was suspended by ropes and leather traces and hoisted by two wooden pulleys. Other materials bought for the engine consisted of beams, hooks, and bolts.[54] At the beginning of the season the engine was assembled, the frame bolted together, and the whole put into working order. During the summer it required frequent overhauling by wheelwrights and carpenters. Every week or two the *proviseurs* purchased more *oint* to grease the engine, so that in 1389 during a campaign of fifty-seven days of pile driving forty-four and a quarter pounds of lubricant were consumed. In September or October the engine was taken apart and stored in the hospital of Saint-Antoine.[55]

During the years 1387–1389 the season for pile driving began in late June. In 1387 the boat to hold the pile driver and sixteen men was described as a *barch*, and when a flood threatened, eight men were required to lift the *barch* and the pile driver out of the Loire. During another season it was a *chaland* which carried the pile driver. Some idea of the size of the *chaland* may be had from the length of two beams, each 5.87 meters long, fastened on either side to protect it.[56] The *chaland*, a freight boat on the Loire, was one of the larger boats at Orleans, but the smaller *sentine* was more often used in connection with work on the bridge, especially for carrying rocks to be

52. A. Tuetey, ed., *Journal d'un bourgeois de Paris* (Paris, 1881), pp. 30–31.
53. Archives du Loiret, CC 920, fol. 27r and passim. See also Gautier, *Traité*, 1:159.
54. At Lyons on 27 February 1421 (n. st.) the consuls moved to recover "un engin et un tour a planter paux" carried off by a flood: Guigue, *Registres*, pp. 288–289. Compare L.F. Salzman, *Building in England down to 1540. A Documentary History* (Oxford, 1962), p. 328.
55. Archives du Loiret, CC 920, fol. 26r.
56. Ibid., fol. 27r.

thrown on the starlings. Large quantities of stone were used. Thus in the week beginning 17 August 1387 thirty-nine boatloads of stone were delivered, in the week of 24 August forty-five, in the week of 31 August forty-six, and in the week of 7 September seventy-two.

When the *chaland* with the engine and sixteen men was ready at the starling, piles were let down from the bridge by rope, and two men in a *sentine* were there to guide the pile to the appropriate spot (see Fig. 1). They held the pile in place while it was being driven and then fastened the piles onto planks to hold them in place at the starling. In three seasons lasting from June to September or October the engine drove an average of about seven to nine piles daily. The men who provided power for the ram earned 20 *d.* each daily, the man in his *sentine* who held the piles 2 *s.* 8 *d.*, his helper 2 *s.*, and a valet who put stone on the starlings 16 *d. p.* daily.[57]

The starling was an idea congenial to medieval bridgebuilders. It required little technical skill and could be used on any but a rock bottom. In addition, there was no question of assembling the large numbers of piles and men necessary for a cofferdam. As for the greater cost of maintaining starlings rather than masonry piers founded on piles, in the twelfth century there does not seem to have been much concern for repairs, and only in the thirteenth century does the problem of methodical maintainence seem gradually to have forced itself on the attention of bridge owners. At all times at most medieval bridges there was a reluctance to make repairs, partly because the required sums were hard to raise. The Pont-Saint-Esprit and the bridge at Orleans were better funded than most.

At least by the fourteenth century founding piers on piles and the use of the cofferdam were common. At Nyons the community in 1361 promised to furnish the contractor with twenty strong men to empty water from the river when it became necessary.[58] In the fifteenth century, as in the eighteenth, it was evidently usual to place a pier on a solid foundation, whether on rock or on dirt reinforced with piles. The consuls of Lyons in 1454 in letting a contract to rebuild a pier specified that the contractor should found it as low and as deep as he could on a good mass of masonry and that if it should be necessary and were advised, he should "drive good and sufficient chestnut piles" under the foundation.[59] Fifty years later the consuls of Lyons were told by masons and carpenters that a foundation should rest on piles a foot and a half

57. Ibid., fol. 25v.
58. Séjourné, *Grandes Voûtes*, 2:29.
59. Guigue, "Notre-Dame de Lyon," p. 252.

in diameter (0.48 m.) and that they should be driven one and a half feet apart as deep as possible and fastened together. Then gravel and quicklime were to be placed to make a platform on which to found the pier.[60]

To drive piles under the place where a foundation was to be located, to make a platform, and to lay up masonry it was necessary to divert the stream or use a cofferdam to ensure dry conditions for working. Details have survived for putting in a cofferdam preparatory to replacing a cutwater of the bridge at Albi in 1408.[61] On Thursday, 16 August 1408, work was begun on demolishing the old cutwater, and it continued Friday and Saturday. Piles measuring 4.96 meters were driven into the riverbed with thinner pieces behind them to form two enclosures, one inside the other, the timbers being fastened at the top by crosspieces. On Monday the men threw *bart* (a clay mortar) between the rows of piling to make them watertight. Tuesday, August 21, the process of exhausting the water inside the cofferdam began. The equipment consisted of buckets made of thin planks by a carpenter, bound with iron rings by a blacksmith, and pulled by ropes. Draining a cofferdam by the use of pails was the method used at Orleans in 1435–1436. It was depicted by Ramelli in 1588, and recommended above all by Hubert Gautier (1765).[62] Gautier especially valued it because, unlike pumps, "baquetage" never broke down from mechanical failure, and if the men changed off every two hours, there was no excessive fatigue. Apparently this was true at Albi, for during the first week six men stayed on the job around the clock for four days, and two did so for five.

At Albi the process of exhausting the water inside the cofferdam went on continuously from Tuesday, August 21, until Saturday, August 25, and from Sunday, August 26, to the following Saturday, September 1. During this time the town provided food for the men to enable them to stay on the job. Probably piles were driven under the foundation of the cutwater, and certainly some laying of masonry took place. When the draining of water from the cutwater was completed, the masons worked another three days; at least part of the time they and their helpers operated from boats.

Albi town accounts also give us details on replacing two arches in 1409–1410. The first requirements were a scaffolding from which the men could work and a temporary wooden bridge to provide passage across the Tarn.

60. Ibid., p. 256, 25 November 1506.
61. M.N. Boyer, "Rebuilding the Bridge at Albi," *Technology and Culture* 7 (1965), 24–37.
62. Capitano Agostino Ramelli, *Le Diverse e Artificiose Macchine* (Paris, 1588), p. 171. Gautier, *Traité*, 1:82.

Neither the carpenter who had contracted to make the centering nor the mason in charge of the brick arch felt it necessary to appear for these preliminaries. Master Arnaut Albert, a carpenter who was "master of the arch," cut wood in the forest to complete the scaffolding and to begin construction of the centering. Fashioning the centering involved no little skill and judgment. First the frame of the centering was assembled on the river bank; then the iron-shod piles were driven and beams laid across them at the spot where the arch was to be. After that the centering was supposed to be lifted into place on the beams, but the centering was found to be too large. In 1409 three days of adjustment were needed before the centering could be raised onto the beams. After the final touches were completed (in the shape of struts and the planks on which the voussoirs were built), the brick arch was laid. Some weeks later the piles were sawed off and the centering removed. The same difficulties recurred in connection with the next arch in 1410. Although as much as possible of the timber of the old centering was used in the new, extensive readjustments were necessary, as the widths of the two arch openings differed. Once again the centering was at first too large, but this time after readjustment it was too small, so that wedges had to be driven underneath it to raise it adequately. Master Arnaut Albert's method of working was evidently one of continual improvisation. Despite the fact that the town had purchased ropes to enable him to measure the arch, he evidently relied very little on them.

The engineering difficulties at Albi illustrate some of the problems of medieval bridge construction. Add to these that on occasion a cofferdam could turn out to be inoperable because of an underground spring, as happened at Orleans during repairs following the floods of 1435.[63] It is then possible to understand some of the reasons why the starling appealed to twelfth-century builders. It required neither expertise nor nice planning, and the initial expense was minor.

In the course of the fifteenth century the accumulated expertise of bridgebuilders had settled the question of the optimum method of founding piers where there was no rock bottom, but trial and error methods had made little progress in decreasing the size of piers. Alberti in the fifteenth century advocated a thickness of one-third of the arch opening, and Gautier in the eighteenth stated that men of former periods had built piers one-third to one-half of the arch openings.[64] Gautier advocated calculating the stresses and

63. Archives du Loiret, CC 967, fol. 16v.

64. Leon Battista Alberti, *Ten Books on Architecture* 8.6, ed. J. Rykwert (New York, 1966), p. 172. Gautier, *Traité*, 1:107.

strains so as to build piers of minimum width, and he asked the mathematician La Hire to provide him with a method for this calculation. The resulting formula left Gautier much disappointed. He complained that La Hire had used algebra, and since no builders knew algebra, the method was useless.[65] (Gautier's book was first published in 1716.) Evidently in the early eighteenth century the theoretical education of bridgebuilders was approximately what it had been in the Middle Ages.

It is notable that the longest arch of the period was supported not by piers but by two solid banks. At Vieille-Brioude, over the Allier, a single segmental arch attained in the fourteenth century the length of 54.23 meters. Begun perhaps in 1340, the arch collapsed in 1822. For nearly 400 years, says Séjourné, it had been the largest arch in the world.[66]

A great pier was largely independent of others, a desirable feature in an age which converted a timber bridge to stone in the course sometimes of decades, sometimes of centuries. The cost of such a pier was not necessarily excessive, inasmuch as often only the facing was of hewn stone, while the interior was rubble or detritus and earth. At Albi clay was stamped down inside a pier by the workmen's feet.[67] Moreover, large piers were essential to the support of houses and towers on the bridge. Evidently to medieval builders the advantages of the huge piers outweighed the disadvantages. By obstructing the freeway, large piers increased the rapidity of the stream, and this accelerated the undermining and overthrowing of the piers. The classic example is Old London Bridge, which blocked a third of the freeway, and at Lyons when the medieval bridge over the Saône was replaced by the July monarchy, the water level dropped so much that a dam was necessary to render navigation upstream feasible.[68] The first piers placed in the Garonne at Agen and in the Rhone at Lyons modified the course of the current. At Lyons, where the piers were eight to twelve meters thick, there were protests in the thirteenth century that on their account the current was washing away the islands.[69] The channel was further constricted by dikes to protect the river banks, and when to these difficulties was added the deep layer of gravel in the Rhone bed, it is no wonder that the water, forced to flow faster and

65. Gautier, *Traité*, 1:21.
66. Séjourné, *Grandes Voûtes*, 2:22.
67. Archives communales d'Albi, CC 180, fol. lxxxviii verso. G.C. Home, *Old London Bridge* (London, 1931), p. 34.
68. L. Bruguier-Roure, *Les Constructeurs de ponts au moyen-âge* (Paris, 1875), p. 36, n. 1.
69. Guigue, "Notre-Dame de Lyon," p. 194.

deeper, washed the gravel and piles from under the piers and again and again toppled them. Contemporaries, however, ascribed the difficulties of the builders of the Pont-de-la-Guillotière to the "harshness" of the waters.[70]

In the eighteenth century the bridge at Orleans was pulled down after 600 years of service and replaced with a modern structure by Perronet.[71] A comparison of the bridge built in 1750–1760 with its predecessor affords enlightening contrasts between the architectural and engineering ideas of the two periods. Eighteenth-century opulence and technical competence allowed Perronet to decide that his bridge must at all costs achieve harmonious proportions. Arch openings of approximately equal width and piers of the same size meant the location of piers without reference to the nature of the bottom. The site is an awkward one, for the bed of the Loire at this point is sandy and full of springs, at the same time that the river is subject to freshets of redoubtable force. Perronet used two regiments of soldiers to empty water from the inside of cofferdams, and even so their efforts were more than once nearly overwhelmed by the river. Gautier had stated that piers should be founded on rock or on piles, and piles should be driven until they refused the ram. At Orleans in 1750–1760 the workmen again and again lost piles without reaching bottom: one pier, supposedly completed, began to settle and required so much more work that it was nicknamed the "Golden Pier" because of the expense.

It was all very unmedieval. Perronet even removed the sandy island, the Motte Saint-Antoine, where the hospital had been situated and which had been such a boon to the medieval builders because it decreased the number of piers to be founded in water. Medieval parsimony would not have involved builders in such an exorbitant cost as the Golden Pier. The essential practicality of the bridge constructors of the Middle Ages was indeed of a parochial, even a short-sighted nature, but it conserved at any one time the rather limited resources of the period.

70. Ibid., pp. 271–272.

71. J.R. Perronet, *Description des projets et de la construction des ponts de Neuilly, de Mantes, d'Orléans et d'autres* (Paris, 1782–1783), 2:1 ff.

11

The Versatility of Medieval Bridges

EARLIER ages took a view of bridges different from that of the present day. According to one modern dictionary, a bridge is merely "a structure erected over a depression or obstacle . . . , carrying a roadway for passengers, vehicles, etc." This limited definition would have been inadequate in former times. Leon Battista Alberti (1404–1472), writing in Renaissance Italy from the point of view of an architect and town planner, considered the bridge, along with the "Cross-Ways and Place for public Spectacles," as among the parts of the street principally to be adorned.[1]

> The Bridge, no doubt, is a main Part of the Street; nor is every Part of the City proper for a Bridge; for besides that it is inconvenient to place it in a remote corner of the Town, where it can be of Use to few, and that it ought to be in the very Heart of the City, to lie at hand for everybody; it ought certainly to be contrived in a Place where it may easily be erected and without too great an Expence, and where it is likely to be the most durable.[2]

In the Middle Ages in France there certainly would have been agreement that the bridge should be centrally located, but in other ways the medieval bridge was distinctive. Not only did it ordinarily lack the symmetry and stone decorations of the Renaissance bridge, but the medieval bridge attracted other buildings. It was very far from being confined to use as a roadway for passengers and vehicles, although that was almost always the overriding consideration. It was surrounded and covered by towers, chapels, hospitals, houses, shops, markets, mills, and fishponds.

1. Alberti, *Ten Books on Architecture*, trans. C. Bartoli and J. Leoni, ed. J. Rykwert (New York, 1966), p. 172 (trans. first published in 1726).
2. Ibid., p. 76.

To wish to live on a bridge was a typically medieval idea—there were houses on a Parisian bridge as early as the sixth century—and one which lasted into the eighteenth century. Not until that time were the houses on the Pont-Notre-Dame at Paris, on the Pont-de-Tarn at Albi, and on London Bridge demolished. In the Middle Ages bridges were thoroughly integrated into town life. To reside on a bridge meant that one missed as little as possible of what was going on.

In the medieval use of the expression "roads and bridges" the latter were incomparably the more important. It had not been so in ancient times. The Romans had given the priority to roads; and it is Roman roads that have aroused the admiration of recent centuries, while their bridges have excited comparatively little interest, this despite a number of excellent surviving examples. Modern attention has been focused on Roman roads because of their impressive construction and the immensity of the undertaking, providing an empire with a means of communication. Medieval values were different, not only in France but in England as well. It has been written that "as roads were the great contribution of the Romans" to the British system of internal communications, "so the substitution of bridges, often substantially built, for fords was the great public work" of the medieval period.[3] In Gaul the Romans showed little enterprise in the permanent bridging of the larger rivers or in the construction of stone bridges. The solitary Roman bridge over the lower Rhone was built of boats, and stone bridges over the Seine were constructed only in the Middle Ages.

The Gauls seem to have placed a higher value on bridges than had the Romans, as shown by the Gallic penchant for naming localities "Briva," bridge. Medieval people in France continued this habit. The Roman Rotomagus became Pont-de-Ruan (Indre-et-Loire), the Castrum Seicum of 889 became the Ponts-de-Cé of the fifteenth century,[4] and the village of Saint-Saturnin-du-port (Gard), where the first stone of the Pont-Saint-Esprit was laid in 1265, by the fifteenth century had become Pont-Saint-Esprit. Towns, hamlets, manors, fiefs like the Pont-en-Vertais (Loire-Inférieure), a chapel like the

3. C.T. Flower, ed., *Public Works in Medieval Law*, 2 (London, 1923), xix.
4. A. Longnon, *Géographie de la Gaule au VIe siècle* (Paris, 1878), p. 286. For the change from Arioca to Pontarlier, see M. Vivien de St. Martin, *Nouveau Dictionnaire de géographie universelle* (Paris, 1879–1900), s.v. Pont-de-Gennes (Sarthe) was "burgus de Gena" in 1235 but Le-Pont-de-Gene in 1357. H. Gröhler, *Über Ursprung und Bedeutung der französischen Ortsnamen*, 2 (Heidelberg, 1933), 147.

Pont-Pierre (Yonne), commune Villeblevin, and a convent like the Pont-aux-Dames (Seine-et-Marne)[5] were all named after bridges. Even geographical features like swamps and brooks took their names from bridges.[6] The tendency to consider the presence of a bridge the most noteworthy feature of a locality seems to have increased, especially during the later Middle Ages. In a list of bridges attested between 1000 and 1500, 40% of those mentioned in the eleventh, twelfth, and thirteenth centuries had places named after them, but in the fourteenth century the proportion rose to 50% and in the fifteenth century to 60%.

The Merovingians, on conquering Roman Gaul, brought with them the German idea of customary law. Thus only existing bridges had the right to repairs by the inhabitants of the region. This conception should have inhibited bridge construction on new sites, and throughout the Middle Ages there was continual insistence on precedent. Nothing, however, could prevent the period from being one of profound change. For example, during the Middle Ages the orientation of the road system was altered. Agrippa had constructed in Gaul four new roads centering in Lyons, but in the high Middle Ages most roads led to Paris, the capital of the kings of France. There was a new north-south orientation which demanded river crossings at unaccustomed sites. Medieval people in France insisted that, wherever feasible, these crossings should not be by ferries or fords but by bridges.

In general, medieval bridges prospered during periods of strong central government and of an expanding economy. Merovingian and Carolingian rule was not conducive to bridge construction and there was difficulty even in maintaining existing spans. Still it was the tenth century that represented the nadir in construction. The disappearance of central power meant that bridges were appropriated by local magnates who seized other parts of the imperial power, such as the right to collect certain taxes. In the tenth and eleventh

5. In the commune of Saix (Vienne) a farm was called "Le Pont" in 1476. The Pont-en-Vertais (Loire-Inférieure) mentioned in the fourteenth century is now within the city limits of Nantes (*Dic. top. Loire-Inférieure,* s.v.). The priory called Saint-Jerosme in 1079 or 1080 was referred to as Pont-aux-Moines (Loiret), com. Mardie, in 1119 (*Layettes*, 1:41–42). Even streets were named after bridges, like the Rue du Pont-Gras in Louviers (Eure) in 1455 (*Dic. top. Eure*, s.v.).

6. In the commune of Béthune (Pas-de-Calais) there was an old swamp called Pont-a-Vaches in 1215 (*Dic. top. Pas-de-Calais*, s.v.). In the arrondissement of Pau (Basses-Pyrenees) in 1277 there were *landes* named after Le Pont-Long (*Dic. top. Basses-Pyrénées*, s.v.), and elsewhere there were brooks called after bridges, as the Pont-Rouge (Isère), com. Allemont, in the fourteenth century (*Dic. top. Isère*, s.v.) and Pontauriol (Aude), com. La Cassaigne, in 1468 (*Dic. top. Aud*, s.v.).

centuries a fragmented and anarchic Aquitaine originated the Peace of God movement. At this time (the eleventh century) also appeared the idea of the bridge as a pious work, so that bridge construction merited a heavenly reward. Concomittantly there arose the legal entity of the bridge corporation, the endowed bridge, and also the bridgebuilding brotherhoods. The *opus pontis* seems to have occurred in France only in those regions affected by the Peace of God movement. The *fratres pontis* appeared at only three sites in southeastern France.

The concept of the bridge as a public benefit was stated at Albi in the early eleventh century, and it has been suggested that the idea never died out in the South. In any case, it is only in southern sources (including, of course, Lyons) that these researches have uncovered statements on the bridge as a "public utility" or "common benefit." Such declarations are rare. Documentary evidence from all parts of France indicate that the bridgebuilder's motives were defense, accessibility, and hopes for a spiritual or financial reward.

After the disappearance of the Carolingian *corvée* the earliest evidence for taxes for bridge support comes from southern France, the same area which mobilized the resources of charity for this purpose. Yet in southern as well as northern France there is considerable evidence that bridges were treated by many as private property. The dominance of local lords, lay and ecclesiastical, in bridge construction in the eleventh and twelfth centuries gradually gave way with the rise of the communes, so that the tendency in the last centuries of the Middle Ages in France was for more and more towns to acquire responsibility for regional bridges. The growth of governmental regulation and provision coincided with the increase in power of the Capetian monarchy. In the course of the fourteenth and fifteenth centuries bridges were widely recognized as a governmental responsibility to be financed chiefly by taxes on inhabitants of the region. At this time, it even occasionally happened that bridge construction, a matter of local initiative since the tenth century, was begun at the insistence of the central government. The tempo of bridgebuilding in France after 1000 can be summarized from the appendix: 66 in the eleventh century, 119 in the twelfth, 202 in the thirteenth, 165 in the fourteenth, and 96 in the fifteenth.

A recognition of the role of trade, population levels, and governmental authority in bridge construction should not lead one to neglect geography. The two provinces of medieval France in which most bridges are attested after 1000 are Périgord (the department of Dordogne) with thirty-four and

Dauphiny (the departments of Drôme and Isère) with eighty-one. Both are in mountainous regions, and in Dauphiny many streams are torrents, making ferries extremely hazardous. Hence there was a particular inducement for bridge construction.

The buildings on a medieval bridge expressed varying aspects of its influence on the life of the times. For example, towers on or at the entrance to bridges were typical of their military role in the Middle Ages. The importance of bridges in warfare was known to the Romans; Caesar's object in building his famous bridge across the Rhine was to move his army across the river into Germany. The idea of bridges as important to troop movements continued, of course, in the Middle Ages, but the later period gave a novel twist to the strategic importance of bridges. From the ninth century on bridges were fortified and garrisoned to deny to the enemy passage not only over the bridge but also along the river under it. Charles the Bald seems to have originated this use of bridges, and where the defenders displayed the requisite determination, it was very successful.

Medieval fortifications began to come into their own only toward the end of the tenth century. The concept of the fortified bridge was uncongenial to the age in which the Peace of God movement flourished. When at the end of the tenth century and during the eleventh regional assemblies were called to bring moral pleasure to bear on recalcitrants to observe the rights of the church and of the poor, part of the program was directed against stone castles. There were other and later protests, as when in 1296 the pope insisted on the destruction of three towers, one in the middle and one at each end, of the Pont-Saint-Esprit.[7] On the whole, efforts to prevent or even to limit the construction of fortifications for bridges were ineffectual. Thus in the fourteenth century the Pont-Saint-Esprit was fortified.[8] The building of a bridge tower by the lord was one of the first means of countering resistance by townsmen. Two towers at the bridge across the Saône at Lyons figured in an agreement ending a struggle between the bishop and chapter on the one hand and the citizens on the other. One tower was to be destroyed and the other placed in the guard of Eudes, duke of Burgundy.[9] The Hundred Years' war provoked a surge in military building. Part of the strategy of Charles V was to

7. L. Bruguier-Roure, ed., *Cartulaire de l'oeuvre des église, maison, pont et hôpitaux du Saint-Esprit*, *Mém. Acad. Nîmes*, ser. 7, vol. 17 (1894), Annexe, p. xxxix.

8. Viollet-le-Duc, *Dictionnaire raisonné de l'architecture française* (Paris, 1854–1868), 7:228.

9. V. Mortet, P. Deschamps, eds., *Recueil de textes relatifs à l'histoire de l'architecture, XIIe–XIIIe siècles* (Paris, 1929), p. 210.

fortify all towns as refuges for the inhabitants of the countryside so as to deny supplies to an invading army. Bridges were an important part of communal defenses. At Albi in the second half of the fourteenth century the bridge tower, which had fallen from neglect, was rebuilt.[10] Fortified bridges figured in many a military action. When the French and Bretons took Châtellerault in 1371, the chief English officer hastily fled in the middle of the night to safety in the bridge tower.[11] During the Armagnac-Burgundian troubles each new group to seize Paris garrisoned the Pont-de-Charenton.[12]

Chapels and hospitals built on bridges were a visible symbol of their spiritual role, a characterization which goes back to the eleventh century. A charitable institution, it was appropriate for a bridge to own a chapel, support a priest, and supervise a hospital. There were only a handful of cases in which the administration of a hospital and a bridge were combined, and in only two of these does the association seem to have endured and been successful—at Orleans and at Metz. At Romans there was a lying-in hospital on a pier, and at Orleans the hospital was located on the island crossed by the bridge. Chapels continued to be built on bridges into the eighteenth century, and throughout the Middle Ages they were important to the image of the bridge as a pious work and as a means of encouraging donations to bridges. Often an episcopal or papal indulgence would specify that the spiritual benefits of donating money or labor to a bridge could become effective only after the donor had confessed and been absolved in the bridge chapel. The fact that almost no bridge chapels were located north of the Loire in France reminds the reader that in France the idea of the bridge as a pious work was primarily a southern idea.

Offerings for bridge maintenance were received both in chapels and at Orleans and at Blois in alms trunks placed in front of the chapel. Charitable solicitations for bridge support continued from the late twelfth century to the end of the Middle Ages, and in the early part of the period at a number of sites it was spectacularly successful. Yet enthusiasm waned, and donations decreased, so that although superior generosity supplied larger sums for the alms trunks of the bridge over the Rhone at Lyons, it did not make a significant

10. A. Vidal, ed., *Douze comptes consulaires d'Albi du quatorzième siècle* (Albi, 1906), p. 59.
11. Froissart, *Oeuvres*, ed. K. de Lettenhove (Brussels, 1870–1877), 7:463.
12. Georges Chastellain, *Oeuvres*, ed. K. de Lettenhove (Brussels, 1863–1866), 1:130. P. Champion, *Vie de Charles d'Orléans* (Paris, 1911), pp. 116–117. According to the Ordonnance Cabochienne, there was a captain of the bridge of Saintes: *L'Ordonnance cabochienne 26–27 mai 1413*, ed. A. Coville (Paris, 1891), p. 30.

contribution to bridge expenses in the early fifteenth century. Beginning in the twelfth century, and to a large extent in the thirteenth, bridgebuilders turned to other sources to finance bridges.

Many a bridge had a toll house connected with it, and at Millau and at Albi the collector's residence was actually on the bridge. At Albi there was a bar across the roadbed to prevent gratuitous passage across the bridge. However, it would be a mistake to consider the toll house one of the most important structures on a medieval French bridge. The contribution of tolls collected for passage across a bridge was not a decisive factor in construction and repair. Throughout the Middle Ages tolls at many a bridge were diverted from its upkeep to other purposes. The notion that tolls were properly a perquisite to be received by some lord or bishop or convent outlasted the Merovingian and Carolingian *corvée.* Then, too, there was the frequent demand that passage across a bridge should be free. This was a corollary of the concept of the bridge as a pious work. The charter of Count Eudes II of Blois declared that since he must not be deprived of his heavenly reward for bridgebuilding, all persons should cross his bridge at Tours gratuitously.[13] Yet by the late twelfth century most bridges had acquired tolls, even including that at Tours. The demand of the communes was also for free passage, but only for their own citizens. In the late twelfth century the town of Agen had requested of Richard I the right to collect alms for bridges; after the middle of the thirteenth century towns were petitioning the king for permission to collect tolls and taxes.

A toll house must have been conveniently located for collecting charges on boats passing under the bridge. At bridges like Pont-Saint-Esprit and at Orleans where passage across the bridge was free (except on market days at Orleans), passage under it was not. Even so, at Orleans the chief reliance of the *opus pontis* was not on tolls but on rents from houses and land. In the middle of the thirteenth century at a time of increasing demand for bridges, steps were taken to make tolls more effective in financing them. Thus the Parlement of Paris decreed that the recipient of tolls was responsible for the upkeep of a bridge. A number of abbeys then divested themselves of ownership, and more and more towns found themselves responsible for bridges. If tolls were not very successful in financing bridges, collection of charges at barriers (*barrages*) and of excise taxes on salt and wine was much more effective for this purpose. In the fourteenth century courts held the inhabitants of

13. E. Martène, U. Durand, *Thesaurus novus anecdotorum* (Paris, 1717), 1:176.

a region liable for repairs to bridges. Also, the royal government by way of the courts not only enforced the rules that owners of bridges should keep them in repair; on occasion it took the novel step of insisting that local people build new bridges.

The agglomeration of buildings on and at a medieval bridge not only served the military and spiritual needs of the neighborhood, but also provided everyday services. Houses and shops were built on bridges from Merovingian times. There were houses or shops or both on bridges at Abbeville, Agen, Albi, Angers, Espalion, La Rochelle, Orleans, Paris, Toulouse, Tours, and elsewhere. It is clear that owners and *pontonarii* or *operarii* who managed the fabric encouraged the construction of houses on the bridge because of the income. This was true at La Rochelle in the early thirteenth century; and at Blois, when in the sixteenth century the bridge was having financial difficulties, it was suggested that more houses be constructed on it so as to provide rent.[14]

One might have supposed that the danger from floods would have acted as a deterrent to the construction of houses on bridges. For example, in 1206 the three arches of the Petit Pont in Paris with their houses were swept away.[15] Nevertheless, floods failed to discourage people preferring to live on bridges. Despite the numerous times the Petit Pont was swept away in the twelfth and thirteenth centuries, the acts of the cathedral chapter continue to record the sale of houses or the transfer of the *cens* to be paid on this or that dwelling.

In the twelfth and thirteenth centuries there were masters of the University of Paris living and teaching on the bridges of Paris. It was widely agreed that bridges were a good place to buy or sell or to collect alms. Leper hospitals desired the privilege of placing "unum bacinum supra Magnum Pontem" at Paris.[16] As for shops on bridges, Joinville in his life of St. Louis mentions a conflagration in the business section of Damietta. It was just the same, he wrote, as if one had fired the Petit Pont.[17] It is known that there were jewelers, drapers, and moneychangers on the Grand Pont with the row of jewelers opposite to that of the moneychangers. Because of the latter the name Pont-

14. A. de Martonne, "Notice historique sur l'ancienne pont de Blois," *Mém. Orléanais* 6 (1863), 441.
15. *Historiens de France*, 18:798.
16. J. Imbert, *Histoire des hôpitaux français* (Paris, 1947), p. 197, n. 1.
17. Joinville, *Histoire de Saint Louis*, ed. N. de Wailly, 12th ed. (Paris, n.d.), p. 69 (ch. 24, #164). H. Sauval, *Histoire et recherches des antiquités de la ville de Paris* (Paris, 1724), 1:216.

au-Change replaced the former name of the bridge. It does not seem to have been usual to have masons' and carpenters' workshops on the bridge, but in 1318–1321 the town of Limoux promised to pay rent on fifty such shops on the bridge to the king in return for permission to collect a *barragium* on passersby to finance a new bridge.[18] A site on a bridge was obviously advantageous to a butcher shop, as at Blois.[19]

At small towns and provincial centers houses were often only scattered along bridges. This was true at Orleans and Albi, for example, but at Toulouse and Paris bridges were so completely covered with houses as to merit the name "covered bridge." At Toulouse the Pont-de-la-Daurade was so-called because of the half-timbered houses on it, and the Pont-Notre-Dame at Paris, begun in 1413, had a double row of houses, all of the same height.[20] Congestion of buildings on the bridges of Paris developed at least by the thirteenth century, as shown by the small size of some of the shops and workshops on the Grand Pont. At both the Grand Pont and Petit Pont the demand for space meant that buildings extended far back from the roadway, supported on piles.

An occasional bridge possessed a mill house on the roadbed but very many had mills associated with them, preferrably downstream but also sometimes upstream. At a bridge of any size there were commonly several mills. This was true of bridges at Paris, at Orleans, at Blois, Joigny, Pont-sur-Yonne, and elsewhere. There were mills at bridges, and there were mill bridges, that is bridges built to afford access to mills. More than one mill next a Parisian bridge had its own arch connecting it with the bank so that grain and flour could be easily transported. There was even a famous millers' bridge, the Pont-aux-Meuniers, at Paris, which served eleven mills. There were various technological reasons for placing a mill beneath a bridge, among them that of incorporating parts of the bridge into the mill race. However, the compelling reason for the location was probably that the same population which crossed the bridge made use of the mill to grind its grain.

There were also fishponds beneath bridges. Sometimes the reason was that it was very easy to convert an existing millpond underneath a bridge into a pen by merely enclosing the upstream end, since the two sides and the

18. C. Devic, J. Vaissète, *Histoire générale de Languedoc* (Toulouse, 1872–1893), 10: 592–594.

19. *Cartulaire de la ville de Blois (1196–1493)*, ed. J. Soyer and G. Trouillard (Paris, 1907), p. 109.

20. J. de Lahondès, *Les Monuments de Toulouse; histoire, archéologie, beaux-arts* (Toulouse, 1920), p. 317. G. Corrozot, *Les Antiquitez, chroniques, et singularitez de Paris*, 2nd ed. (Paris, 1561), p. 149r.

sluice gate were already there. In a number of cases the gift of a site for a mill at a bridge included the right to establish a fishpond. This was true at Beaugency, Pontoise, and at Poitiers. The abbey of Jumièges (996?–1026) was given three mills and the fishponds installed between the arches of the bridge at Pont-de-l'Arche.[21]

A medieval bridge in France with its accumulation of buildings and its arches of varying widths represented the willingness of people of that period to improvise. They adapted the bridge to its environment by founding the piers where the bottom was suitable, and in the twelfth century they conserved their slender resources by placing piers on starlings to avoid any great expenditure at one time. They used segmental and pointed arches as well as semicircular arches, and frequently they constructed pointed cutwaters downstream as well as upstream.

It is obvious from the surviving examples that medieval builders were capable of producing bridges that were artistically pleasing in their balanced proportions, but they did not set the same value on symmetry as did the eighteenth century. Dissimilar cathedral towers were able to evoke admiration on the part of medieval people in France, and so also were asymmetrical bridges. Medieval men did indeed consider their bridges beautiful, but their idea of decorating them did not consist in attaching pilasters to the piers but in building attractive houses on the bridge or in hanging tapestries along the roadbed on ceremonial occasions. Frequently the harmonious relationship between the piers and arch openings was obscured above by the superstructure of houses, chapels, and towers, and below by watermills and their sluices. That medieval people admired and valued their bridges is well illustrated by the accounts for the reconstruction of the bridge at Orleans after the flood of 1435: "Donation of a person who had great love and affection for the bridge and its rebuilding."[22]

Alberti wrote, "The Bridge, no doubt, is a main Part of the street,"[23] and this idea was shared by people in medieval France. Their bridges, cluttered with houses, stores, workshops, towers, chapels, mills, and other buildings, constituted a center of town life. In such a case to live at or on a bridge was to be at the heart of things. At various times and places in the Middle Ages it was possible to be born on a French bridge, to attend scholarly lectures on a

21. E. Lot, "Mélanges carolingiens. Suite II, Le Pont de Pîtres," *Le Moyen-Age* 18 (1905), 24, n. 1.
22. A. Collin, *Le Pont des Tourelles à Orléans* (Orelans, 1895), p. 495.
23. Alberti, *Ten Books*, p. 76.

bridge, to earn one's living and to purchase one's supplies on the bridge, to attend services and receive the sacraments there, and to spend one's old age in a hospital for the aged. (On death it was necessary to leave the bridge, for although St. Bénézet was inhumed on a bridge at Avignon, such burial must have been rare in the extreme.)

The multiple uses of bridges was distinctively a medieval concept. It was the product of an age in which ideas and functions were combined rather than separated, in which natural science was a part of philosophy and in which the aim of history was to instruct. In the Middle Ages bishops frequently had temporal as well as spiritual powers, and royal government employed both clergy and laymen in administrative posts. The medieval bridge with its agglomeration of buildings was the expression of a period, in which, at least in northern France, many individuals had rights in the same piece of land, and in which a bridge served a variety of purposes.

Appendix

List of Medieval French Bridges with Dates when First Attested

The following list is arranged alphabetically according to the location of the bridge. If it is known, the name of the bridge is given, followed by the earliest date when there is evidence for the existence of the bridge. The bibliographical references consist of a Roman numeral referring to the appropriate section of the bibliography, the alphabetical entry within that section, and volume and page numbers.

Abbeville (Somme). Pont-aux-Poissons. 1207. II: Prarond, 48.

Agen (Lot-et-Garonne). Fund collection authorized 1189. Passable by end 13th cent. IV: Tholin, 440.

Airvault (Deux-Sèvres). 11th cent. IV: Emerson and Gromort, 15.

Albi (Tarn). Ca. 1035. IV: *Gallia Christiana*, 1, Instrumenta, 4.

Amboise (Indre-et-Loire). 1110. II: "Ex gestis amb.," 510.

Amiens (Somme). Pont de la Bretesque. 1415. III: Amiens, 1:AA, p. 2.

Andelys, Les (Eure). 1197. II: Mortet and Deschamps, 175.

Angers (Maine-et-Loire). 1028. IV: Bienvenue, 213–214.

Angoulême (Charente). Pont Saint-Cybard and pons sancti Eporchii. 1144–1149. II: Nanglard, 161.

Arc-en-Barrois (Haute-Marne). 1157. IV: Gröhler, 2:149.

Arcs-de-Parigny, Les (Vienne). 980. IV: Gröhler, 2:149.

Arc-sur-Tille (Côte-d'Or), cant. Dijon-est. 1124. IV: Gröhler, 2:148.

Arles (Bouches-du-Rhône). Roman bridge of boats after 1st cent., extant in 4th and 5th cents. but not in medieval times. IV: Brogan, 32.

Ars-sur-Moselle (Moselle), cant. Gorge. 881 (Arx), 889 (Arcus). IV: Gröhler, 2:149.

Art-sur-Meurthe (Meurthe-et-Moselle). 770 (Arcus). IV: Gröhler, 2:149.

Auboiré, *see* Metz.

Aurillac (Cantal). 1298. Pont dal Boys, on the Jordanne. IV: Champollion-Figeac, 107.

Ausque (Pas-de-Calais), near Thérouanne, cant. Aire-sur-Lys, arr. Saint-Omer. Bridge over the Hem. Before 1178. II: Mortet, *Recueil*, 392.

Autun (Saône-et-Loire). Ancient and 1253. II: Charmasse, 180–181. For Pont-Saint-Andoche, IV: Fontenay, 204.
Auvers-sur-Oise (Oise), arr. Pontoise. Ancient, also 885. IV: Lot, 1.
Auxance (Vienne), com. Chasseneuil. III: Vienne, s.v.
Auxerre (Yonne). Ancient and before 1075. IV: Quantin, 431.
Avignon (Vaulcuse). Pont-Saint-Bénézet. 1177–1188. IV: Lefort, "Le Premier Pont," 387–392.
Balme, La (Isère), com. St.-Paul-d'Izeaux. 14th and 15th cent. IV: Sclafert, 383.
Balme-en-Genevois, La (Haute-Savoie), arr. Annecy. 1370. II: Mugnier, 415.
Baudemont (Marne), cant. Anglure, arr. Epernay. 1203. Two bridges. II: Mortet and Deschamps, 196–197.
Bayonne (Basses-Pyrénées). Pont Maior, 1298. Pont de Bertaco, 1381. II: Ducéré, 96, 216.
Beaugency (Loiret). Before 1160–1182. II: Vignat, 87, 124.
Beaumont (Yonne), cant. Seignelay. Before 1185. II: Quantin, 2:359.
Beaumont-sur-Oise (Seine-et-Oise). 1143–1145. II: Depoin, 2:#285 ter.
Beaumont-sur-Sarthe (Sarthe), arr. Mamers. 12th cent. IV: Bienvenue, 456.
Beauvais (Oise). Before 1122. IV: Loisel, 266.
Bergerac (Dordogne). 1209. III: Dordogne, s.v.
Besançon (Doubs). Ancient and medieval. IV: Marnotte, 121 ff.
Béziers (Hérault). Before 1209. IV: Viollet-le-Duc, 7:229.
Bigaroque (Dordogne), sec. de la com. Coux. 1243. III: Dordogne, s.v.
Blanzy (Saône-et-Loire), com. Blanzy. 1305. II: Charmasse, 3:127.
Blau (Aude). 1259. IV: Aude, *Inv.*, 4:99.
Blois (Loire-et-Cher). Before 1078. IV: De Martonne, 415, 421.
Bognens, Le pont de (Ain), com. Andart-Condom. 1290. III: Ain, s.v.
Boisseron (Hérault). Roman. II: Dartein, 1:5.
Bonevi (Seine-et-Marne), near Lorrez-le-Bocage. 1202. IV: Lot and Fawtier, clxviii.
Bonpas (Vaucluse). 1166. IV: Michel, 370.
Bordeaux (Gironde). 1292. II: Delpit, 1–2.
Boudelin (Aube), com. Fontaine. 1231. III: Aube, s.v.
Bourges (Cher). Bridge to vineyard. 1095. II: Mortet, *Recueil*, 103, n. 2.
Bouveruel (Seine-Maritime), in bailliage of Rouen. 1334. II: Delisle, *Actes*, 67, 68, 73.
Bouxières-aux-Dames (Meurthe), cant. Nancy-Est. 1073. IV: Calmet, 1, Preuves, 474.
Branthôme (Dordogne), arr. Périgueux. Pont du Monastère. 1474. III: Dordogne, s.v.
Breuilpont (Eure). 1336. IV: Gröhler, 2:127.
Bréval (Seine-et-Oise), cant. Bounières. Bridges rebuilt in provost's district. 1202. IV: Lot and Fawtier, cxlv, cxlvii.
Brèves (Nièvre). On the Yonne. Gallo-Roman. IV: Grenier, 2:264.

Brézolles (Eure-et-Loire). Pontchartrain in Brézolles on the Nouvette. Gallo-Roman. IV: Grenier, 2:262.
Briare (Loiret). On the Loire. Gallo-Roman. IV: Grenier, 2:264.
Brières (Ardennes). Gallo-Roman. IV: Grenier, 2:264.
Brieulles-sur-Bar (Ardennes). Gallo-Roman. IV: Grenier, 2:264.
Brieulles-sur-Meuse (Meuse). Called Briodurum in 11th-12th cent. IV: Grenier, 2:264.
Brimeux (Pas-de-Calais). 1292. IV: Champollion-Figeac, 105.
Brioude (Haute-Loire). Gallo-Roman. IV: Grenier, 2:264.
Brioux (Deux-Sèvres). Gallo-Roman. IV: Grenier, 2:264.
Briovera, later Saint-Lo (Manche). Gallo-Roman. IV: Grenier, 2:264.
Brissac (Maine-et-Loire). After 1162. IV: Bienvenue, 236.
Brissarthe (Maine-et-Loire), cant. Châteauneuf-sur-Sarthe. Gallo-Roman. IV: Longnon, *Les Noms*, 47, #98.
Brives (Indre). Gallo-Roman. IV: Grenier, 2:264.
Brives (Mayenne). Gallo-Roman. IV: Grenier, 2:264.
Brives-la-Gaillarde (Corrèze). Gallo-Roman. 6th cent. IV: Grenier, 2:264.
Brives-sur-Charente (Charente-Inférieure). Gallo-Roman. IV: Grenier, 2:264.
Brivodura, *see* Pontaubert (Yonne).
Brivodurum, *see* Pont-Audemer (Eure).
Bugue, Le (Dordogne), arr. Sarlat. On the Vézère. 1463. III: Dordogne, s.v.
Cahors (Lot). Roman bridge. Pont Neuf, 1251. Pont Valentré, 1308–ca. 1355. IV: Dartein, 1:7 ff.
Cajarc (Lot). 1222. IV: Collin, *Le Pont*, 370–371.
Calais (Pas-de-Calais). 1390. II: Smith, 14.
Campagnac (Gard), com. Sainte-Anastasie. Pont-de-Saint-Nicolas de Campagnac. 1261. IV: Germer, 185.
Carbonne (Haute-Garonne). Rebuilding, 1356. II: *Ordonnances*, 3:82.
Carcassone (Aude). 1184. II: Mahul, 5:313.
Castres (Tarn). Pont Bielh and Pont de Navès. Repairs, 1391. III: Castres, BB, 2.
Céret (Pyrenées-Orientales). 1321–1339. IV: Séjourné, 1:10 ff.
Chabris (Indre). Gallo-Roman. IV: Grenier, 2:264.
Chalonnes-sur-Loire (Maine-et-Loire), arr. Angers. 1138–1148. IV: Bienvenue, 218.
Châlons-sur-Marne (Marne). 1164. II:Mortet and Deschamps, 114.
Chalon-sur-Saône (Saône-et-Loire). Roman bridge of timber on stone piers. Pont Saint-Vincent built of stone 1415 on ancient piers. IV: Bartier, 129.
Charenton (Seine). Gallo-Roman and 7th cent. II: Leroux, 226, n. 1.
Charité-sur-Loire, La (Nièvre). 1482. IV: Girardot, 24.
Chartres (Eure-et-Loir). Pont de Falaise in the parish of Saint Prest. 1374–1454. III: Eure-et-Loir, *Inv.*, Ser. G., 6:22.
Chartroussas (Drôme), cant. Guignan, com. Roussac. 1334. III: Drôme, *Inv.*, Ser. E., 4:255.

Château-Gaillard, *see* Andelys, Les.
Château-Gontier (Mayenne). 1080. III: Mayenne, *Dic. Hist.*, 1:577.
Châteauneuf-de-Mazenc (Drôme), cant. Dieulefit. 1462. III: Drôme, *Inv.*, Ser. E., 4:138, E 6341.
Châteauneuf-sur-Sarthe (Maine-et-Loire). 1136–1138. II: Urseau, 286–287.
Châteauthierry (Aisne). 1287. II: Longon, 3:66.
Châtellerault (Vienne). Ca. 1060. IV: Crozet, 508.
Chaudefonds (Maine-et-Loire), on the Layon, near Chalonnes-sur-Loire. 1451. II:Lecoy de la Marche, 149–150.
Chauvigny (Vienne). Ca. 1080. IV: Crozet, 507.
Cheney (Yonne), cant. Seignelay. Before 1185. II: Quantin, *Cartulaire*, 2:357–358, #cccxliv.
Chérisy (Seine-et-Loise), near Argenteuil. 1266. II: Boutaric, 1:100, #1087.
Chinon (Indre-et-Loire). 1115–1126. IV: Chartrou, 387 ff.
Chocques (Pas-de-Calais), arr. Béthune. Castle bridge. 1170–1191. II: Mortet and Deschamps, 118.
Choisy-au-Bac (Oise), cant. and arr. Compiègne. 1248. II: Mortet and Deschamps, 256.
Claix, *see* Grenoble.
Clamecy (Nièvre). 1147. IV: Quantin, 425.
Coëmont (Sarthe), cant. Château-du-Loir, arr. Mans. 11th cent. IV: Bienvenue, 459.
Commines (Nord), cant. Quesnoy-sur-Deule, arr. Lille. 2nd half 14th cent. II: Froissart, 10:110.
Compiègne (Oise). Ca. 876. Also 1112. II: Morel, 1:64.
Comporté (Vienne), communes Saint-Macou and Saint-Saviol. Pont de Compourte. 1432. III: Vienne, s.v.
Condom (Gers). Rebuilt 1281. II: Bémont, 2:125, #461.
Conflans (Savoie). Pont de Conflans, on the Arly. Peutinger Table. IV: Desjardins, 4:154.
Confolens (Charente). On the Vienne. Ca. 1110–1140. IV: Crozet, 503.
Conques (Aude). Pont de Vilar. 1083. II: Desjardins, 57.
Corbeil (Seine-et-Oise). 1301. II: Viard, 750.
Cornil (Corrèze), cant. and arr. Tulle. 1103. II: Mortet, *Recueil*, 15–16.
Corre, La (Aube), com. Champ-sur-Barse. 1215. III: Aube, s.v.
Couilly, *see* Pont-aux-Dames.
Coulanges-sur-Yonne (Yonne), cant. and arr. Auxerre. 12th and 13th cent. IV: Quantin, 425–426.
Coutances (Manche). Over the Vire. 1024–1093. II: Mortet, *Recueil*, 70 ff.
Cravant (Yonne). 1368. II: Delachenel, 2:38.
Creil (Oise). 1415. II: *Ordonnances*, 10:326 ff.
Cuisery (Saône-et-Loire). 1350. On the Seille. II: Jeanton, 119.
Décize (Nièvre). 1225. IV: Blin, 245.

Dijon (Côte-d'Or). Pont Arnault. Wooden bridge on the Ouche. 1395–1396. III: Dijon, 1, Ser. B., 24–25.
Donchéry (Ardennes). 1322. II: Saige, 1:633.
Dormans (Marne), arr. Epernay. 1258–1259. II: Bourquelot, 72.
Douve (Manche), com. Carentan. 1346. IV: Denifle, 1:36.
Durtal (Maine-et-Loire), arr. Angers. 11th cent. IV: Bienvenue, 460.
Entraygues (Aveyron). Pont Notre-Dame, on the Lot, 1269. Bridge on the Truyère, perhaps 14th cent. IV: Dartein, 1:43 ff.
Escaupont (Nord). Peutinger Table, Pons Scaldis. Scalpons, 921. IV: Desjardins, 4:127.
Espalion (Aveyron). 1060. II: Desjardins, 401 ff.
Estaing (Haute-Loire), com. Monastier. 1329. III: Haute-Loire, *Dic.*, s.v.
Etampes (Seine-et-Oise). 1060–1108. II: Fliche, 126.
Evreux (Eure). Pont Saint. 1245. III: Eure, s.v.
Fénestrange (Meurthe), arr. Sarrebourg. Pons Saravi. Peutinger Table. IV: Desjardins, 4:54, 133.
Ferté-Milon, La (Aisne), arr. Château-Thierry, cant. Neully-Saint-Front. 1349. IV: Champollion-Figeac, 137.
Figeac (Lot). 1291. IV: Champollion-Figeac, 137.
Flogny (Yonne), arr. Tonnerre. 1224. II: Quantin, *Recueil*, 138, #316.
Florent (Marne). Pont-de-Rêmes. Gallo-Roman. IV: Grenier, 1:174.
Fontaine-Daniel, Abbey (Mayenne), com. Saint-Georges-Buttavent. Bridge over the Mayenne, 1209. Pont-David, 1231. II: Gross-Duperon, 71–72, 143.
Fontaine-le-Port (Seine-et-Marne). 1204–1205. II: Depoin, 3, #626.
Fouchères (Aube), com. Bar-sur-Seine. 1220. III: Aube, s.v.
Frégère, *see* Najac.
Gaillac (Tarn). 1256. IV: Rossignol, 580.
Gers. Two bridges over the river Gers. Ca. 1060. II: Mortet, *Recueil*, 260 ff.
Gien (Loiret). 1458. II: Lecoy de la Marche, 154.
Gières (Isère), com. and cant. Grenoble-Sud. Pont de Gières. 14th cent. IV: Sclafert, 383.
Gisors (Eure). 1184. II: Mortet and Deschamps, 107–108.
Goncelin (Isère). Pont de Goncelin, on the Isère between the communes of Goncelin and Touvet. 1270. IV: Sclafert, 379–380.
Gournay-sur-Marne (Seine-et-Oise). 1147. II: Mortet and Deschamps, 71.
Grenade-sur-Garonne (Haute-Garonne), arr. Toulouse. 1309. II: Clement, #4344.
Grenoble (Isère). Pont du Drac, near Claix, 1219. IV: M. Champion, 2:195. Pont-Saint-Jaymes, 14th cent. Pont de Porte-Traine, 1378–1379. III: Grenoble, 2:75, CC.
Grillon (Drôme), cant. and com. Grignan. On the road to Grillon. 1436. III: Drôme, *Inv.*, Ser. E, Suppl., 4:222.
Ile-Bouchard (Indre-et-Loire), arr. Chinon. Pont de Saint-Gilles, to Ile-Bouchard in the Vienne. Reconstructed 1451. IV: Lance, 125.

Isle-Adam (Seine-et-Oise), arr. Pontoise. 1415. II: *Ordonnances*, 10:326 ff.

Isle-de-Pons, L' (Vienne), com. Ligugé. On the Clain. 1420. III: Vienne, s.v.

Isle-en-Périgord, *see* Lisle (Dordogne).

Jargeau (Loiret), arr. Orleans and Meung-sur-Loire. 13th cent. II: Mortet and Deschamps, 202.

Jarrie (Isère), cant. Vizille, arr. Grenoble. A bridge at this point briefly about 1310, then a ferry again. IV: Sclafert, 441.

Joigny (Yonne). In existence from 13th cent. IV: Quantin, 437.

Jouars (Seine-et-Oise). Pontchartrain. Gallo-Roman. IV: Grenier, 1:173, n. 4.

Juvardeil (Maine-et-Loire), cant. Châteauneuf-sur-Sarthe, arr. Segré. 1075. II: Marchegay, "Chartes," 399.

Juvisy (Seine-et-Oise), cant. Longjumeau, arr. Corbeil. 13th cent. II: Boileau, 250.

Lagny (Seine-et-Marne). Destroyed 1432. II: Monstrelet, 5:27–28.

Laval (Mayenne). 1170. II: De Broussillon, 1:10.

Lectoure (Gers). 1485. II: Druilhet, 130–131.

Lézigny (Yonne), com. Mailly-la-Ville. 1241. II: Quantin, *Recueil*, 215, #475.

Limoges (Haute-Vienne). Pont-Saint-Martial, rebuilt Roman bridge. Pont-Saint-Etienne. Both in use in 12th cent. IV: Degrand, 2:65.

Limoux (Aude). 1318. IV: Devic and Vaissètte, 10:592–594.

Linde, La (Dordogne), arr. Bergerac. Toll for bridges. 1289. III: Dordogne, s.v.

Lisle (Dordogne), cant. Branthôme. Isle-en-Périgord. 1309. II: *Ordonnances*, 11:417 ff.

Livron-sur-Drôme (Drôme), cant. Loriol. 1471. III: Drôme, *Inv.*, Ser. E, Suppl., 6:235, #E9719 CC8.

Longpont (Aisne), cant. Villers-Cotterets. 1118. III: Aisne, s.v.

Lorrez-le-Bocage, *see* Bonevi.

Lorris (Loiret). 1292. II: Boutaric, 1:442–443, #805.

Louviers (Eure). Ca. 1027. II. Bonnin, 1:6.

Lyons (Rhône). Pont-de-Saône, before 1050. IV: Paradin, 120. Pont-de-la-Guillotière, begun ca. 1180–1182. IV: Guigue, 183 ff.

Mâcon (Saône-et-Loire). Repairs, 1362–1367. II: Michon, 182, n. 1.

Magny-en-Vexin (Seine-et-Oise), arr. Mantes. Two bridges. 1260. II:*Olim*, 1:112.

Maguelone (Hérault). Ca. 1129–ca. 1158. II: Mortet, *Recueil*, 91.

Mailly-la-Ville (Yonne). 14th cent. Later ruined by neglect. IV: Quantin, 426.

Mailly-le-Château (Yonne). 15th cent. IV: Quantin, 426.

Mans, Le (Sarthe). Pont Perrin, 12th cent. Pont Ysoard or Ysoir, 1067–1078, III: Sarthe, s.v.

Mantes (Seine-et-Oise). In 1185 Louis VII confirmed an act of 1121 mentioning the bridge. IV: Luchaire, *Etudes*, 286, #590.

Marcigny (Saône-et-Loire). 12th cent. Later disappeared. IV: Blin, 252, n. 6.

Marquefave (Haute-Garrone), cant. Carbonne, arr. Muret. 1413. II: *Ordonnances*, 10:206.

Marseilles (Bouches-du-Rhône). 1255. Pont des Béroards, 1352. Pont de Velut, 1436. III: Marseilles, *Inv.*, AA, 12. III: Marseilles, *Dic.*, s.v.

Mathefelon (Maine-et-Loire), com. Tiercé, arr. Angers, near Montreuil-sur-Loir. Repaired 1454. IV: Lance, 127.

Mauléon (Basses-Pyrénées). 1289. II: Bémont, 2:435.

Mayenne (Mayenne). 1028. II: Mortet, *Recueil*, 73, n. 2.

Meaux (Seine-et-Marne). 991. II: Richer, 4:50.

Melun (Seine-et-Marne). 1212. II: Mortet and Deschamps, 215.

Mende (Lozère). 14th cent. IV: Dartein, 1:39.

Metz (Moselle). Pont Saint-Georges, before 13th cent. Pont-des-Morts or Moyen Pont, 1222–1223. Pont-Thiffroy, 1222. Le Grand Pont des Morts, 1245. Also at Auboiré near Metz, 1240. IV: Schneider, 13, 18.

Meulan (Seine-et-Oise). 1182. IV: Houth, "Catalogue," 528, #81.

Meung (Loiret), arr. Orleans. 13th cent. II: Mortet and Deschamps, 202.

Mézières (Ardennes). Repairs 1497–1498. III: Godefroy, s.v. Boulevart.

Millançay (Loiret). 1387–1389. I: Loiret, A 1979, fol. 49v.

Millau (Aveyron). Pont Vieux, before 1156. New bridge, before 1286. II: Verlaguet, 356. II: Artières, 27.

Moissac (Tarn-et-Garonne). 1120. I: Bibl. Nat. Doat, CXXXI, fol. 291. IV: Champollion-Figeac, 127.

Montargis (Loiret). 1234. II: De Wailly, 575.

Montauban (Tarn-et-Garonne). Ca. 1336. IV: Didron aîné, 41 ff.

Montélimar (Drôme). 1360. IV: Bruguier-Roure, 60.

Montereau-Faut-Yonne (Seine-et-Marne). Ca. 1016–1037. II: Poupardin, *Recueil Saint-Germain*, 1:86.

Montferrand (Puy-de-Dôme). 1308. III: Clermont-Ferrand, 1:363.

Montier-en-Der (Haute-Marne). 876. II: Poupardin, *Recueil Provence*, 127.

Montignac (Dordogne), arr. Sarlat. Over the Vézère. 11th cent. Stone piers. III: Dordogne, s.v.

Montignac-le-Petit (Dordogne). Over the Ille. 1281. III: Dordogne, s.v.

Montmorot (Jura), suburb of Lons-le-Saulnier. 1439. IV: Gauthier, 314–315.

Montolieu (Aude), arr. Carcassonne. 1392. II: *Ordonnances*, 7:500.

Montouron (Maine-et-Loire). 1458. II: Lecoy de la Marche, 153.

Montpellier (Hérault). Pont sur le Lève. 1267. II: Martène and Durand, 2:461.

Montpont (Dordogne), arr. Riberac. 1376. III: Dordogne, s.v.

Montréal (Gers), arr. Condom. 1411. II: Breuils, 289.

Montréal (Yonne), cant. Isle-au-Serain. Moulin du pont. 1204. II: Quantin, *Receuil*, 18–19, #37.
Montreuil-Bonnin (Vienne), com. Vouillé. On the Boivre. IV: Crozet, 519.
Montreuillon (Nièvre), cant. Château-Chinon. 1213. IV: Quantin, 384.
Montségur (Ariège). 1265. II: Mortet and Deschamps, 291–293.
Moret-sur-Loing (Seine-et-Marne). 1499. IV: Pougeois, 120.
Motte-de-Bourbon (Vienne), com. Pouançay. 1455. II: Lecoy de la Marche, 152.
Mouy (Oise), arr. Clermont. 1320. II: Boutaric, 2:347–348, #6298.
Najac (Aveyron). Pont de la Frégère, on the Aveyron. 1288. IV: Dartein, 1:46–47. IV: Rey, 17.
Nantes (Loire-Inférieure). Pont-en-Vertais. 14th cent. III: Loire-Inférieure, s.v.
Narbonne (Aude). Roman and 1227. II: Sidonius, 1:286. IV: Champollion-Figeac, 152.
Natiaux (Yonne), arr. Auxerre, com. Avrolles. Pont-des-Natiaux, on the Armaçon. 1147. III: Yonne, s.v.
Negueromieu *or* St. Hilaire, *see* Toulouse.
Nevers (Nièvre). 1273. II: Boutaric, 1:177, #195.
Nîmes (Gard). 1308. II: Clement, #2689.
Nogent-sur-Seine (Aube). 1319. II: Longnon, 3:165.
Norges-le-Pont (Côte-d'Or), com. de Norges. 1276. III: Côte-d'Or, s.v.
Nyons (Drôme). 1361. IV: Séjourné, 2:10 ff.
Olla (Isère), com. St. Egrène. 13th cent. III: Isère, s.v.
Orange (Vaucluse). 1368. III: Orange, 1:23.
Orleans (Loiret). Before 1176. IV: Collin, *Le Pont*, 390, 393.
Orthez (Basses-Pyrénées). By 1254. IV: Dartein, 1:29.
Oulchy-le-Château (Aisne), arr. Soissons. 1287, II: Longnon, 3:66.
Palaminy (Haute-Garonne), arr. Muret, cant. Cazères. Washed out 1388. II: Froissart, 11:31.
Pamiers (Ariège). 1356. I: Bibl. Nat. Doat XCIII, fol. 207r ff.
Paris (Seine). Grand Pont, Petit Pont, 1st cent. B.C. II:Caesar, 7.58.5, 6. Pont-aux-Meuniers, after 1296. IV: Fagniez, 160, n. 2. Pont-Saint-Michel, 1378. IV: Sauval, 1:225. Pont-Notre-Dame, 1412–1416. II: Tuetey, 30.
Parrigneux (Ain), near Roussillon. 1290. III: Ain, s.v.
Passavant (Marne), cant. Sainte-Menehould. Drawbridge and Pont-aux-Vendages. 1287. II: Longnon, 3:67.
Périgny-la-Rose (Aube), cant. Villenaux. Bridges of Pugny. 1175. III:Aube, s.v.
Périgueux (Dordogne). Ancient bridges. Stone bridge, 1206. Pons Sancti Hilarii, 1363. III: Dordogne, s.v.
Perpignan (Pyrénées-Orientales). 1275. IV: Douais, 31.
Pessac, Saint-Martin-de- (Gironde), arr. Bordeaux. Bridge at the archbishop's residence at La Motte-de-Pessac. 1354. II: Drouyn, 321–322.

Pichaumeix *or* Pechaumeix (Meuse), com. Saint-Mihiel. 1280–1282. IV: Champollion-Figeac, 104.
Pierre-Châtel (Ain), com. Virginien. Ancient. Replaced 1226. IV: *Gallia*, 15, Instrumenta, col. 320.
Pierrepont (Aisne), cant. Marle. Castrum Petraepontis. 938, 1165. III: Aisne, s.v.
Pierrepont (Calvados), cant. Falaise. Petrepons. 1145. III: Calvados, s.v.
Pierrepont (Oise). 1198. II: Morel, 1:335–336.
Pierrepont-sur-l'Arantèle (Vosges), cant. Bruyères. 1302. III: Vosges, s.v.
Poissy (Seine-et-Oise). 13th cent. II: Boileau, 250.
Poitiers (Vienne). Pont Achard, bridge on the Boivre, 1017. Pont Joubert, 1083. Bridge on the Clain. III: Vienne, s.v. Pont Cyprien, 1101. IV: A. Richard, 1:431. Pontneuf (also called Pont de Rochereuil), to bourg of Montierneuf, 1087. II: Guérin, 9:78. IV: Crozet, 516.
Pomblin (Aube), arr. Troyes, cant. Chaource, com. Loge-Pomblin. Pont Belin. 1287. II: Longnon, 3:194.
Pompertuzat (Haute-Garonne), arr. Villefranche. Jerusalem itinerary. IV: Desjardins, 4:33.
Pompierre (Vosges), cant. Neufchâteau. 6th cent., 1179. III: Vosges, s.v.
Ponant (Isère), com. Livet-et-Gavet. Pons Nahon. 14th cent. III: Isère, s.v.
Ponnesant (Yonne), com. Saint-Martin-sur-Ouanne. Ponsnascentius. 855, 1188. III: Yonne, s.v.
Ponrault (Vienne), com. Migne. Pons Regales. 993–1029. III: Vienne, s.v.
Pontréau (Vienne). 993–1030. IV: A. Richard, 1:214.
Pons (Charente-Inférieure), arr. Saintes. 1242. IV: Gröhler, 2:147. III: Vivien de St. Martin, s.v.
Pons, Le Grand Logis (Isère), com. St.-Pierre-de-Chartreuse. 12th cent. III: Isère, s.v.
Pons Aerarius *or* Ponte Herarum (Bouches-du-Rhône), between Bellegard and Arles. Roman. Jerusalem itinerary. IV: Desjardins, 4:34.
Pons de Craut, *see* Rochy-Condé.
Pons-de-Rastel, Le (Gard), com. Genolhac. 1212. III: Gard, s.v.
Pons Sancti Hugonis (Isère). Bridge on the Bens between the communes of La Chapelle-du-Bard and Arvillard (Savoie). 14th cent. Now called Pont-du-Diable. III: Isère, s.v.
Pont, Le (Aude), com. Peuvert. Pons de Blavo. 1259. III: Aude, *Dic.*, s.v.
Pont (Aveyron). 1484–1489. III: Aveyron, Ser. C., 211.
Pont, Le (Cantal), com. Naucelles. 1442. III: Cantal, s.v.
Pont, Le (Cantal), com. Dionne. 1485. III: Cantal, s.v.
Pont, Le (Cher), com. Feux. 1464. III: Cher, s.v.
Pont, Le (Cher), com. Lugny-Champagne. 1302. III: Cher, s.v.
Pont, Le (Cher), com. Moulins-sur-Yèvre. 1484. III: Cher, s.v.
Pont, Le (Cher), communes of Reigny and Saint-Christophe-le-Chaudery. 1449. III: Cher, s.v.

Pont, Le (Cher), Pralemine or Saint-Priest, domaine, com. Charenton-sur-Cher. Pons Sancti Preijecti. 1319. III: Cher, s.v.

Pont, Le (Côte-d'Or), cant. Auxonne. 937 or 938, 1257. III: Côte-d'Or, s.v.

Pont, Le (Côte-d'Or), com. Bligny-sur-Ouche. In molendino de Ponte. 1238. II: Charmasse, 1:153.

Pont, Le (Côte-d'Or), com. Pont-en-Massène, cant. Semur. 1368. III: Côte-d'Or, s.v.

Pont, Le (Dordogne), com. Grignol. 1308. Molendinum dict. de Ponte, 1308. III: Dordogne, s.v.

Pont, Le (Dordogne), com. Neuvic. 1471. III: Dordogne, s.v.

Pont, Le (Dordogne), com. Vallereuil. 1440. III: Dordogne, s.v.

Pont, Le (Eure-et-Loir), com. Bailleau-sous-Gallardon. Pons Petre. 1241. III: Eure-et-Loir, *Dic.*, s.v.

Pont, Chemin de (Eure-et-Loir). From Vauparfonds to Luisant. 1300. III: Eure-et-Loir, *Dic.*, s.v.

Pont, Le (Gard), com. Tharaux. 1292. III: Gard, s.v.

Pont, Le (Haute-Loire), com. Saint-Pal-de-Mons. 1314. III: Haute-Loire, *Dic.*, s.v.

Pont, Le (Haute-Loire), com. Voûte-Chilheac. 1288. III: Haute-Loire, *Dic.*, s.v.

Pont, Le (Isère), com. Herbeys. Ponte Vitreo. 14th cent. III: Isère, s.v.

Pont, Le (Isère), com. St. Pierre-de-Mésage. Pons de Rippis and Pons de Romanche. 13th cent. III: Isère, s.v.

Pont, Le (Isère), com. Villefontaine. 15th cent. III: Isère, s.v.

Pont, Moulin du (Mayenne), com. Bazouges. 1277. III: Mayenne, *Dic. top.*, s.v.

Pont, Moulin du (Vienne), com. Coussay. 1473. Farm and windmill. III: Vienne, s.v.

Pont, Le (Vienne), com. Genouillé. 1403. III: Vienne, s.v.

Pont, Le (Vienne), com. Lomaizé. 1469. III: Vienne, s.v.

Pont, Moulin du (Vienne), com. Saint-Genest. On the Fontpoise. 1474. III: Vienne, s.v.

Pont, Le (Vienne), com. Saix. Hostel du Pont. 1476. III: Vienne, s.v.

Pont, Le (Vosges), com. Dommartin-lès-Remiremont. 1235. III: Vienne, s.v.

Pont-à-Bucy (Aisne), cant. Crécy-sur-Serre. Pons-de-Nogento-Abbatisse. 1170. III: Aisne, s.v.

Pont-à-Chaussy (Moselle), com. Courcelles-Chaussy. To the left of the Nied Française, 1324. Pont-aux-Loups *or* Pont Quinquobeille, 15th cent. III: Moselle, s.v.

Pont-à-Couleuvre, Le (Aisne), com. Auffrique-et-Nogart. Pons de Colovere. 1145. III:Aisne, s.v.

Pontageon (Haute-Loire), com. Ventenges. Pontago. 1279. III: Haute-Loire, *Dic.*, s.v.

Pontaigon (Vienne), com. Lomaizé. Pontum Aigone, 916. Pontaigum, 1223. III: Vienne, s.v.

Pontailler (Côte-d'Or), arr. Dijon. Pontiliacus, 869. Pons Lindi, 872. In villa Pontiliaco, 1049. III: Côte-d'Or, s.v.
Pontaix (Drôme), com. Die. Mutatio Darentiaca of the Jerusalem itinerary, 333. IV: Desjardins, 35. Pontays, 1215. III: Drôme, *Dic.*, s.v.
Pont-Alain, Le (Mayenne), com. Saint Berthevin. Le moulin et reffoul de Pontalain. 1443. III: Mayenne, *Dic. top.*, s.v.
Pontalery (Calvados), com. united to Mesnil-Durand in 1826. Pont Alerie. 1320. III: Calvados, s.v.
Pont-à-Luc (Gard), communes of Nîmes and Marguerites. Footbridge. 1301. III: Gard, s.v.
Pontamafrey (Savoie), cant. Saint-Jean-de-Maurienne. Pons Amalfredi. 1190. III: Gard, s.v.
Pont Ambroix (Gard). Roman. IV: Desjardins, 4:150.
Pont-à-Mousson (Meurthe). 896, 905, 1230. III: Meurthe, s.v.
Pont-Annet (Morbihan), com. Priziac. On the brook Pont-Rouge. 1431. III: Morbihan, s.v.
Pontarcher (Aisne), com. Ambleny. De Ponte-Archerii. 1320. III: Aisne, s.v.
Pontarcy (Aisne), cant. Vailly. Pons-de-Arserio. 1232. III: Aisne, s.v.
Pontardennes (Pas-de-Calais), com. Wizernes. Pont d'Artengues. 1469. III: Pas-de-Calais, s.v.
Pontarie, La (Dordogne), com. Saint-Laurent. 1460. III: Dordogne, s.v.
Pontarlier (Doubs). 13th cent. On site of Ariorica in the Antonine itinerary. IV: Desjardins, 1:100.
Pontarm (Loire-Inférieure), com. Asserac. De Ponte Armore. 14th cent. III: Loire-Inférierue, s.v.
Pont-Arnaud (Dordogne), com. Monsec. 1373. III: Dordogne, s.v.
Pont-Arnaud (Gard), com. Nîmes. Bridge on the Cadereau. 1380. III: Gard, s.v.
Pontarrane (Aude), com. Compagne. Mill of Pontarrana on the Aude. 1373. III: Aude, *Dic.*, s.v.
Pontarrou, Le (Aude), com. Magrie, sect. A. 1257. III: Aude, *Dic.*, s.v.
Pont-à-Sault (Pas-de-Calais), com. Dourges. 1271. III: Pas-de-Calais, s.v.
Pont-Asquin (Pas-de-Calais), com. Wardecques. 1306. III: Pas-de-Calais, s.v.
Pont-Astier (Haute-Loire), com. Saint-Pal-de-Chalençon. 1163. III: Haute-Loire, *Dic.*, s.v.
Pontaubert (Yonne), cant. Avallon. Pons-Herberti. 1167. III: Yonne, s.v.
Pont-Audemer (Eure). Brivodurum, in antiquity. Duos Pontes, 715. Pont-Audemer, 1027. IV: Delisle and Passy, 2:548.
Pont-Auffroy (Seine-et-Marne), com. Chevru. Mill. 1210. III: Seine-et-Marne, s.v.
Pontaujard (Drôme), com. Montbrison. Pons Aujart. 1332. III: Drôme, *Dic.*, s.v.
Pontault (Eure-et-Loir), com. Nottonville. 1468. III: Eure-et-Loir, *Dic*, s.v.
Pontauriol (Aude), com. Cassaigne. A brook. 1468. III: Aude, *Dic.*, s.v.

Pont-Authou (Eure), com. and cant. Montfort-sur-Riscle. Pons Altou. 1024. III: Eure, s.v.

Pont-Auvray, Le (Mayenne), com. Laudivy. 12th cent. III: Mayenne, *Dic. top.*, s.v.

Pont-aux-Dames (Seine-et-Marne), cant. Crécy, com. Couilly. 1226. II: *Layettes*, 2:79–80.

Pont-aux-Moines (Loiret), com. Chéry and Mardie, cant. Orleans. 1075. II: Prou, 192.

Pont-à-Vaches (Pas-de-Calais), com. Béthune. 1215. III: Pas-de-Calais, s.v.

Pont-à-Vendin (Pas-de-Calais), com. Lons. 1024. III: Pas-de-Calais, s.v.

Pontavery (Aisne), cant. Neufchâtel. Pons Vardius. 1112. III: Aisne, s.v.

Pontazel (Aude), com. Bram. 1315. III: Aude, *Dic.*, s.v.

Pontbagnes (Isère). Formerly near Izeaux. 13th cent. III: Isère, s.v.

Pont-Barrois, Le (Cher), com. Comeressault. 1345. III: Cher, s.v.

Pont-Barse (Aube), com. Courteranges. 1183. III: Aube, s.v.

Pontbeau (Savoie). A domain. 12th cent. III: Savoie, s.v.

Pont-Bellenger (Calvados), cant. Saint-Sever. 1203. III: Calvados, s.v.

Pont-Benoist (Seine-et-Marne), com. Sainte-Colombe. Mill and old fief. 1256. III: Seine-et-Marne, s.v.

Pont-Bernard, Le (Côte-d'Or), com. Montmaçon. 1303. III: Côte-d'Or, s.v.

Pont-Bernard (Isère), between communes of Beaufin and Aspres-les-Corps. On the Drac. 14th cent. III: Isère, s.v.

Pont-Besnier, Le (Sarthe), between communes of Juigne and Ausvas-le-Hamon. 1321. III: Sarthe, s.v.

Pont-Besse, Le (Cantal), com. Laveissiere. 1490. III: Cantal, s.v.

Pont-Bone (Dordogne), com. Lembras. 1373. III: Dordogne, s.v.

Pontcallec (Morbihan), com. Berne. Brook and pond. 1291. III: Morbihan, s.v.

Pontcarlet (Isère), com. Hueg. 14th cent. III: Isère, s.v.

Pont-Carpin (Isère), communes of Grenoble and Saint-Martin-d'Hères. On the brook Grande-Morgne. 13th cent. III: Isère, s.v.

Pontcarré (Seine-et-Marne), cant. Tournon. Fief. 15th cent. III: Seine-et-Marne, s.v.

Pontcerme (Aude), com. Coursan. A farm. Pons Septimus, 1352. III: Aude, *Dic.*, s.v.

Pont-Cervier, Le (Haute-Loire), com. Cohade. On the Vendage. 1323. III: Haute-Loire, s.v.

Pontcharain (Isère), com. Theys. Pons Charen. 13th cent. III: Isère, s.v.

Pontcharra (Isère and Savoie), com. St. Honoré. 13th cent. III: Isère, s.v.

Pontchartrain (Eure-et-Loir), *see* Brézolles.

Pontchartrain (Orne), com. Saint-Mard-de-Réno. Gallo-Roman. IV: Grenier, 1:173, n. 4; 2:262.

Pontchartrain (Seine-et-Oise), *see* Jouars.

Pontchâteau (Loire-Inférieure), arr. Saint-Nazaire. 11th cent. III: Loire-Inférieure, s.v.

Pont-Chauvet, Le (Cher), com. Celle-Condé. 1450. III: Cher, s.v.
Pont-Croissant (Isère), com. Montbonnet-Saint-Martin. 14th cent. III: Isère, s.v.
Pont-d'Aby, Le (Aude), com. Narbonne. Old bridge on the road from Narbonne to Capestrang. 1329. Versus pontem David. III: Aude, *Dic.*, s.v.
Pont-d'Ain (Ain), arr. Bourg. 1326. III: Ain, s.v. IV: Gröhler, 1:354, gives the date as 13th cent.
Pont-d'Ainon, Le (Cher), com. Plaimpied-Givaudins. 1300. III: Cher, s.v.
Pont-d'Aisy, Le (Côte-d'Or), com. Aisy-sous-Thil. 1255. III: Côte-d'Or, s.v.
Pont-d'Ambel (Isère), between communes of Ambel and Corps. On the Drac. 13th cent. III: Isère, s.v.
Pont-Dandon (Gard), cant. Molières. 1301. III: Gard, s.v.
Pont-d'Anglars (Dordogne). 1489. III: Dordogne, s.v.
Pont-d'Aurel, Le (Drôme), communes of Aurel and Vercheny. 1193. III: Drôme, *Dic.*, s.v.
Pont-d'Authie (Cantal), com. Jussac. 1369. III: Cantal, s.v.
Pont-d'Avene (Hérault), com. Gigean. 1376–1378. III: Hérault, s.v.
Pont-d'Avord, Le (Cher), com. Farges-en-Septaine. 1460. III: Cher, s.v.
Pont-de-Barangeon, Le (Cher), com. Vignoux-sur-Barangeon. 1260. III: Cher, s.v.
Pont-de-Barret, Le (Drôme), com. Dieulefit. Locus qui prius dictus est Savenna et modo dicitur ad Pontem. 956. III: Drôme, *Dic.*, s.v.
Pont-de-Beauvoisin (Isère), arr. La Tour-du-Pin. Pontem Castellum. 11th cent. III: Isère, s.v.
Pont-de-Brion (Isère), communes Lavars and Roissard. 14th cent. III: Isère, s.v.
Pont-de-Briques, Le (Pas-de-Calais), com. Saint-Leonard. 1203. III: Pas-de-Calais, s.v.
Pont-de-Camarès (Aveyron). 1311. II: Verlaguet, 468.
Pont-de-Cernon (Isère), com. Chapareillan. 14th cent. III: Isère, s.v.
Pont-de-Champ (Isère). Bridge over the Romanche between the communes of Jarrie and Champ. 13th cent. III: Isère, s.v.
Pont-de-Chargy, Le (Cher), com. Barneçon. 1350. III: Cher, s.v.
Pont-de-Charréas. Charrées (Haute-Loire), com. Malrevers. 1253. III: Haute-Loire, *Dic.*, s.v.
Pont-de-Chéruy (Isère), cant. Meyzieu. 13th cent. III: Isère, s.v.
Pont-de-Cognet (Isère). Bridge over the Drac between communes of Cognet and Saint-Jean-d'Hérans. 13th cent. III: Isère, s.v.
Pont-de-Coly (Dordogne), cant. Montignac. 1460. III: Dordogne, s.v.
Pont de Cordéac et de Quet-en-Beaumont (Isère). Above the village of Gautiers. Pons Rumieuf. 13th cent. III: Isère, s.v.
Pont-de-Coubon (Haute-Loire). Coubon, cant. Sud-est du Puy. 1095. III: Haute-Loire, *Dic.*, s.v.

Pont-de-Cros, Le (Cantal), com. Murat. Pons del Cros apud Muratum. 1289. III: Cantal, s.v.
Pont-de-Gavet (Isère), com. Livre-du-Genet, near Grenoble. Gallo-Roman. IV: Desjardins, 4:155.
Pont-de-Gennes (Sarthe), cant. Montfort-le-Rotrou. Le Pont de Gene. 1357. III: Sarthe, s.v.
Pont-de-Gers (Isère), com. Vienne. 15th cent. III: Isère, s.v.
Pont-de-Jaillon, Le (Meurthe), com. Jaillon. Le Pont-a-Jaillons. 1291. III: Meurthe, s.v.
Pont-de-la-Beaurone (Dordogne), com. Chancelade. 1115. III: Dordogne, s.v.
Pont-de-la-Bourne (Isère), com. Villard-de-Lans. 14th cent. III: Isère, s.v.
Pont-de-la-Chartreuse, Le (Haute-Loire), com. Brives-Charensac. Parvus Pons. Ca. 1210. III: Haute-Loire, *Dic.*, s.v.
Pont-de-la-Morge, Le (Isère), com. Moirans. Bridge on the Morge. 14th cent. III: Isère, s.v.
Pont-de-la-Motte (Isère). On the Drac between communes of Avignonet and La Motte-Saint-Martin. 13th cent. III: Isère, s.v.
Pont-de-la-Peyre, Le (Aude), com. Fanjeaux. 1442. III: Aude, *Dic.*, s.v.
Pont-de-la-Peyre (Dordogne), com. Bergerac. 1467. III: Dordogne, s.v.
Pont-de-la-Raz (Isère), com. Valjouffrey. 14th cent. III: Isère, s.v.
Pont-de-l'Arche (Eure), arr. Louviers. Ca. 1020. III: Eure, s.v.
Pont-de-la-Reynette (Gard), cant. Nîmes. On the Fontaine. 1380. III: Gard, s.v.
Pont-de-la-Romanche (Isère), com. Bourg d'Oisans. Pons Medius. 13th cent. III: Isère, s.v.
Pont-de-l'Enceinte, Le (Haute-Loire), com. Yssingeaux. Mill and bridge on the Lignon. Pons voc. de la Saynta. 1273. III: Haute-Loire, *Dic.*, s.v.
Pont-de-l'Escure, Le (Cantal), com. Saint-Flour. La Plancha. 1369. III: Cantal, s.v.
Pont-de-l'Herbasse, Le (Drôme), com. Clérieu. 1468. III: Drôme, *Dic.*, s.v.
Pont-de-Limon (Seine-et-Marne), com. Courtomer, at Pontpierre. 1365. III: Seine-et-Marne, s.v.
Pont-de-Livier, Le (Gard), com. Bellegarde. 1380. III: Gard, s.v.
Pont-de-l'Oula, Le (Isère), com. St. Pierre de Mésage. 14th cent. III: Isère, s.v.
Pont-de-Lussac, Le (Vienne), com. Mazerolles. 12th cent. IV: Crozet, 506.
Pont-de-Maillé, Le (Vienne), com. Saint-Martin-Lars. 1482. III: Vienne, s.v.
Pont-de-Marne, Le (Haute-Marne), com. Langres. 854. III: Haute-Marne, s.v.
Pont-de-Mars, Le (Haute-Loire), com. Chambon. 1254. III: Haute-Loire, *Dic.*, s.v.
Pont-de-Moeurs, Le (Marne), communes Moeurs and Sézanne. Le moulin du Pont-à-Meure. 1493. III: Marne, s.v.
Pont-de-Montgnon (Seine-et-Marne), com. Juilly. 1467. III: Seine-et-Marne, s.v.

Pont-d'Entraygues, Le, *see* Entraygues.
Pont-de-Pierre, Le (Cher), com. Morogues. 1457. III: Cher, s.v.
Pont-de-Pierre, Le (Côte-d'Or), com. Auxonne. 1442. III: Côte-d'Or, s.v.
Pont-de-Pierre, Le (Vosges), cant. Bellefontaine. 1436. III: Vosges, s.v.
Pont-de-Portes, Le (Isère). Ruined bridge on the Romanche, communes Séchilienne and Saint-Barthélemy-de-Séchilienne. 14th cent. III: Isère, s.v.
Pont-de-Quaix (Isère), com. Quaix. Bridge on the Vence. 14th cent. III: Isère, s.v.
Pont-de-Quart, Le (Drôme), com. Beaumont-lès-Valence. 1483. III: Drôme, *Dic.*, s.v.
Pont-de-Rochefort, Le (Haute-Loire), com. Alleyras. Bridge on the Allier. 1345. III: Haute-Loire, *Dic.*, s.v.
Pont-de-Roison, Le (Isère). Bridge between communes of Nantes-en-Ratier and Siévox. 13th cent. III: Isère, s.v.
Pont-de-Ruan (Indre-et-Loire), arr. Tours, cant. Montbazon. 6th cent. The ancient Rotomagus of the Peutinger Table. Also 9th cent. IV: Longnon, 286.
Pont-de-Saint-Martin, Le (Dordogne), com. Agonac. 1367. III: Dordogne, s.v.
Pont-de-Saint-Martin, Le (Isère). Bridge over the Orbanne between communes of Clelles and Saint-Martin-des-Clelles. 14th cent. III: Isère, s.v.
Pont-de-Saint-Nicolas de Campagnac, *see* Campagnac.
Pont-des-Anjalvis (Cantal), com. Menet. 1441. III: Cantal, s.v.
Pont-des-Echelles (Isère). Bridge on the Guiers-Vif, between communes of Entre-deux-Guiers and Les Echelles. 13th cent. III: Isère, s.v.
Pont-de-Soirans (Soirans-Fouffrans) (Côte-d'Or), arr. Dijon, cant. Auxonnes. 1390–1391. IV: Gauthier, 298.
Pont-d'Estrouilhas, Le (Haute-Loire), coms. of Aiguilhe and Espaly-Saint-Marcel. On the Bourne. 1245. III: Haute-Loire, *Dic.*, s.v.
Pont-de-Sumène (Haute-Loire), com. Blavozy. 1359. III: Haute-Loire, *Dic.*, s.v.
Pont-d'Etaules (Côte-d'Or), com. Saint-Seine-sur-Vingienne. 1322. III: Côte-d'Or, s.v.
Pont-de-Tréboul, Le (Cantal), com. Sainte-Marie. 15th cent. III: Cantal, s.v.
Pont-de-Tugny, Le (Aisne), com. Tugny. 1197. III: Aisne, s.v.
Pont-de-Vaux (Ain), arr. Bourg. Ca. 1250. III: Ain, s.v.
Pont-de-Vaux (Saône-et-Loire), com. Marly-sous-Issy. 1312. II: Charmasse, 3:143.
Pont-de-Vaux (Vienne), com. Millac. 1451. III: Vienne, s.v.
Pont-de-Veyle (Ain), arr. Bourg. 1186. III: Ain, s.v.
Pont-de-Veyton, Le (Isère). Bridge on the brook Veyton between the communes of Allevard and Pinsot. 12th cent. III: Isère, s.v.
Pont-d'Isalguier, *see* Toulouse.
Pont-d'Iverny, Le (Sarthe), com. Montmirail. Ca. 969. III: Sarthe, s.v.

Pont-d'Ognon, Le (Aude), com. Homps. 1231. III: Aude, *Dic.*, s.v.
Pont-d'Ouilly (Calvados), com. Saint-Marc-d'Ouilly. 1125. III: Calvados, s.v.
Pont-du-Boeuf (Isère), com. St. Blaise-du-Buis. 14th cent. Pons Boum; Pons de Bos. III: Isère, s.v.
Pont-du-Bourniou, Le (Cantal), com. Ladinhac. 1464. III: Cantal, s.v.
Pont-du-Château (Puy-de-Dôme). Before 1151. II: Suger, 123.
Pont-du-Diable, *see* Pons Sancti Hugonis.
Pont-du-Fieu, Le (Haute-Loire), com. Saint-Julien d'Ance. On the Ance. 1417. III: Haute-Loire, *Dic.*, s.v.
Pont-du-Fort, Le (Calvados), com. Troarn. 1310. III: Calvados, s.v.
Pont-du-Gard (Gard), com. Remoulins. On the Gardon. Roman aqueduct. Called Pons de Gartio in 1295. III: Gard, s.v.
Pont-du-Gy, Le (Pas-de-Calais), communes of Duisans and Etrun. Le pons Duisy. 1364. III: Pas-de-Calais, s.v.
Pont-du-Monastère, *see* Branthôme (Dordogne).
Pont-du-Prat, Le (Isère), com. Valjouffrey. On the Bonne. 14th cent. III: Isère, s.v.
Pont-du-Prêtre, Le (Isère), com. Valbonnais. Pons Sacerdoto. 15th cent. III: Isère, s.v.
Pont-du-Roi, Le (Seine-et-Oise), com. Tigery. Ancient Roman. No evidence for Middle Ages. IV: Grenier, 2:263.
Pont-du-Sault. Bridge over the Rhone between the communes of Porcien Amblagnieu (Isère) and Villebois (Ain). 14th cent. III: Isère, s.v.
Pont-du-Vau, Le (Côte-d'Or), com. Til-Châtel. Molendinum Pontis de Valle. 1203. III: Côte-d'Or, s.v.
Pont-du-Vers (Isère), com. St. Chef. 15th cent. III: Isère, s.v.
Pont-Echanfré (Eure), cant. Broglie. First name of Notre-Dame-du-Hamel. Pons Herchenfret in 11th cent., but Beata Marie de Hamilla in 1260. III: Eure, s.v.
Pontécoulant (Calvados), cant. Condé-sur-Noireau. Pons Escoulandi. 11th cent. II: Calvados, s.v.
Pontempeyrat (Haute-Loire), com. Craponne-sur-Argon. Pons Empeyra. 1311. III: Haute-Loire, *Dic.*, s.v.
Pont-en-Royans (Isère), arr. St. Marcellin. 11th cent. III: Isère, s.v.
Pont-en-Vertais (Loire-Inférieure). 14th cent. Now inside Nantes. III: Loire-Inférieure, s.v.
Pont-Érembourg (Calvados), com. Saint-Denis-de-Méré. 1108. III: Calvados, s.v.
Pont-Evêque (Isère), cant. Vienne-Nord. 11th cent. III: Isère, s.v.
Pontevin (Haute-Marne), com. Châtenay-Mâcheron. Terra de Pontguayn, 1219. Pont Vahin, 1240. III: Haute-Marne, s.v.
Pont-Eyraud (Dordogne), com. and cant. Saint-Aulaye. 1245. III: Dordogne, s.v.
Pont-Farcy (Calvados), cant. Saint-Sever. 1278. III: Calvados, s.v.

Pontfaverger (Marne), cant. Suippes. Pons Fabricatus. Beginning of 11th cent. III: Marne, s.v.
Pontfère (Cantal), com. Laveissière. 15th cent. III: Cantal, s.v.
Pont Flavien (Bouches-du-Rhône), near Saint-Chamas. Roman. Extant. IV: Grenier, 1:568.
Pontfol (Aude), com. Saint-Michel-de-Lanès. 1353. III: Aude, *Dic.*, s.v.
Pontfrault (Seine-et-Marne), com. Château-Landon. Pons Feraudi. Ca. 1090. III: Seine-et-Marne, s.v.
Pontgallet (Eure-et-Loire), com. Bazoche-Gouet. Pons-Galeti. 1215. III: Eure-et-Loir, *Dic.*, s.v.
Pont-Gibaut (Cher), com. Saint-Christophe de Chaudry. Place name. 1178. III: Cher, s.v.
Pontgibert (Haute-Loire), com. Saint-Berain. Pons Gilbert. 1343. III: Haute-Loire, *Dic.*, s.v.
Pont-Gilbert (Seine-et-Marne), com. Torcy. Molendinum de Ponte Gilbert. 1263. III: Seine-et-Marne, s.v.
Pont-Givart (Aisne), communes Arainville and Pignicourt. Aumenancourt-le-Grand and Aumenancourt-le-Petit. 1146. III: Aisne, s.v.
Pontgouin (Eure-et-Loir), cant. Courville. Pons Godonis. 1099. III: Eure-et-Loir, *Dic.*, s.v.
Pont-Gras, Rue du (Eure), arr. Louviers, in Louviers. 1455. III: Eure, s.v.
Pont-Gros (Isère), com. Tullins. 14th cent. III: Isère, s.v.
Pont-Hannois (Seine-et-Marne), com. La Croix-en-Brie. 1270. III: Seine-et-Marne, s.v.
Ponthaut (Isère), com. Allevard. Nemus Pontis Altis. 14th cent. III: Isère, s.v.
Ponthaut (Isère), communes. St. Laurent-en-Beaumont and Sousville. 10th and 13th cents. III: Isère, s.v.
Pont-Hébert (Eure). Quarter of the fief Meuille. 1411. III: Eure, s.v.
Pont-Hémery, Le (Côte-d'Or), com. Brazey-en-Plaine. Pons Haymerici. 1225. III: Côte-d'Or, s.v.
Pont-Herbert (Manche), com. Saint-Clair. 1361–1388. II: Izarn, 51.
Ponthévrard (Seine-et-Oise). Pons Ebrardi. 1162. IV: Gröhler, 2:312.
Ponthion (Marne), cant. Thiéblemont. Pontico, ca. 590. Pons Ugone, ca. 768. Ponteum, 1150. III: Marne, s.v.
Pont-Honfroy (Seine-Maritime), com. Rouen. 4th cent., ca. 1050. Place name has disappeared. IV: Adigard, 320.
Ponthouion (Sarthe), cant. Marolles-les-Braults. 11th cent. III: Sarthe, s.v.
Pont-Hubault (Sarthe), com. Ruardin. 1270. III: Sarthe, s.v.
Pont-Hubert, Le (Aube), com. Pont-Sainte-Marie. 1154. III: Aube, s.v.
Pontifol (Calvados), com. joined in part to Victot-Pontifol and in part to Authieux-sur-Corbon. Ponsfolli, Pons Stulti. 1297. III: Calvados, s.v.
Pontignol, Le (Aude), com. Fanjeaux. Ad Pontilhol, 1345. Ad Pontem de Insula, 1364. III: Aude, *Dic.*, s.v.
Pontignol, Le (Aude), com. Fonters-du-Razès. 1347. III: Aude, *Dic.*, s.v.

Pontignol, Le (Aude), com. Laurac. 1330. III: Aude, *Dic.*, s.v.
Pontignol, Le (Aude), com. Narbonne. 1157. III: Aude, *Dic.*, s.v.
Pontignol, Le (Aude), com. Villasavary. 1139. III: Aude, *Dic.*, s.v.
Pontigny (Moselle), com. Condé sur le Nied. 1339. Pont-deniet, 1404. III: Moselle, s.v.
Pontigny (Yonne), arr. Auxerre. Before 1140 a new bridge. II: Quantin, *Cartulaire*, 1:344, #CCV.
Pont-Jehan (Loire-Inférieure), com. Saint-Lyphard. 1413. III: Loire-Inférieure, s.v.
Pont Julien (Vaucluse), near Apt. Roman. IV: Emerson and Gromort, Plate IV.
Pont-l'Abbé (Manche), com. Saint-Mère-l'Eglise. 1369. II: Izarn, 20.
Pont-Landry, Le (Mayenne), com., cant. Ambrières, arr. Mayenne. 1209. II: Grosse-Duperon, 71–72.
Pont-la-Ville (Haute-Marne), com. Châteauvillain. 1196. III: Haute-Marne, s.v.
Pont-lès-Bonfays (Vosges), cant. Darney. 10th cent., 1303. III: Vosges, s.v.
Pont-l'Evêque (Calvados). 12th cent. III: Calvados, s.v.
Pont-l'Evêque (Finistère). 12th cent. IV: Gröhler, 2:390–391.
Pontlevoy (Loir-et-Cher). Pons leviatus. 1075. IV: Gröhler, 2:147.
Pontlieue (Sarthe), suburb of Le Mans. Bridge on the Huisne. Pontileuga, 616. Ad pontem Leuge, 10th cent., 1212. III: Sarthe, s.v.
Pont-Long, Le (Basses-Pyrénées), arr. Pau. 1277. III: Basses-Pyrénées, s.v.
Pontloup (Seine-et-Marne), com. Moret-sur-Loing. Old priory of Saint-Pierre, dependent on the abbey of Vézelay. Pontloe, ca. 1150. Pons Luppae, 1285. III: Seine-et-Marne, s.v.
Pontmain (Mayenne), com. Saint-Ellier. De Ponte Monii. 1225. III: Mayenne, *Dic. top.*, s.v.
Pont-Marès, Le (Gard), com. Saint-André-de-Valborgne. 1437. III: Gard, s.v.
Pontmarsa (Aude), com. Villasavary. 1332. III: Aude, *Dic.*, s.v.
Pont-Minard (Haute-Marne), com. Forcey. 1174. III: Haute-Marne, s.v.
Pontmoulin (Seine-et-Marne), com. Chailly-en-Brie. Molendinum de Pont Morlen. 1132. III: Seine-et-Marne, s.v.
Pont-Neuf, Le (Drôme), com. Die. Pont Fract, on the Drôme. 13th cent. III: Drôme, *Dic.*, s.v.
Pont-Neuf, Le (Haute-Loire), com. Polignac. Le Pont-Nou. 1408. III: Haute-Loire, *Dic.*, s.v.
Pontoise (Oise, Seine-et-Oise). Two Gallo-Roman bridges called Briva Isarae. 10th cent., 1227. IV: Longnon, *Les Noms*, 168, #700. IV: Champion, 2:159. IV: Champollion-Figeac, 152.
Pontôme (Sarthe), com. Courgains. 1314. III: Sarthe, s.v.
Pontorson (Manche). Repairs 1366. II: *Ordonnances*, 4:636.

Pontoux (Saône-et-Loire). Pons Dubis. Peutinger Table. IV: Gröhler, 2:147.
Pont-Pastoul (Dordogne). Bridge on the Drôme. 12th cent. III: Dordogne, s.v.
Pont-Pérant (Isère), com. St. Laurent-du-Pont. Ruined bridge on the Guiers-Mort. 14th cent. III: Isère, s.v.
Pontpérier (Aude), com. Laurac. 1432–1519. III: Aude, *Dic.*, s.v.
Pont-Perrin, Mill of (Eure), com. and cant. Evreux, at Plessis-Grohan. 1187. III: Eure, s.v.
Pont-Perrin, Le (Vienne), com. Pairoux. On the Clain. 1404. III: Vienne, s.v.
Pont-Peyrat (Dordogne). 1403. III: Dordogne, s.v.
Pont-Peyrène (Bouches-du-Rhône), arr. Marseille, com. Roquesvaire. On the Huveaune. 1452. III: Marseille, *Dic.*, s.v.
Pontpierre (Mayenne), com. Désertines. 1220. III: Mayenne, *Dic. top.*, s.v.
Pontpierre (Moselle), cant. Faulquemont. To the right of the Nied Allemande. 1332. III: Moselle, s.v.
Pont-Pierre (Seine-et-Marne), communes Bernay and Courtomer. On the Yères. Ca. 1210. III: Seine-et-Marne, s.v.
Pont-Pierre (Yonne), com. Villeblevin. 1290. III: Yonne, s.v.
Pontpoint (Oise), arr. Senlis, cant. Pont-Saint-Maxence. 1199. II: Depoin, 3:165, #564.
Pontréan (Ile-et-Vilaine). Preserves name of Roman bridge. IV: Grenier, 2:263.
Pont-Renard (Haute-Loire), com. Saint-Pal-de-Chalençon. 12th cent. III: Haute-Loire, *Dic.*, s.v.
Pont-Rigault (Vienne), com. Trois-Moutiers. 1416. III: Vienne, s.v.
Pont-Robert (Dordogne), com. Bergerac. 1460. III: Dordogne, s.v.
Pont-Roide (Seine-et-Marne), com. Meaux. On the Marne. Farinarium super fluvium Maternam ad Pontem rapidum. 9th cent. III: Seine-et-Marne, s.v.
Pont-Rouft, Le (Gard), com. Aiguesvives. Ad Pontem-Ruptum, ad Pontem-Fractem. 1299. III: Gard, s.v.
Pont-Rouge (Isère), com. Allemont. Bridge on the Romanche. 14th cent. III: Isère, s.v.
Pont-Rougier (Haute-Loire), com. Chomalix. On the Arzon. 1404. Bridge destroyed. III: Haute-Loire, *Dic.*, s.v.
Pont-Roumieu (Dordogne), com. Saint-Germain-et-Mons. 1290. III: Dordogne, s.v.
Pont-Rousseau (Loire-Inférieure), com. Rezé. 1135. III: Loire-Inférieure, s.v.
Pont-Roux (Charente), com. Rouillac. 1075–1101. II: Nanglard, 40.
Pont-Royal (Isère), com. Chapareillan. On the Glandon. 14th cent. III: Isère, s.v.
Ponts, Les Cinq (Dordogne), com. Meuvic. A fief. 1490. III: Dordogne, s.v.

Ponts (Seine-Inférieure), arr. Dieppe, com. Eu, cant. Ponts-et-Marais. 1181–1189. II: Berger, 7,2:383.
Pont-Sainte-Maxence (Oise). 779. IV: Gröhler, 2:420.
Pont-Saint-Esprit (Gard). Begun 1265. II: Bruguier-Roure, 3 ff.
Pont-Saint-Laurent (Isère), com. Saint-Laurent-du-Pont. Bridge on the Guiers-Mort. 15th cent. III: Isère, s.v.
Pont-Saint-Manut (Dordogne), com. Douville. 1310. III: Dordogne, s.v.
Pont-Saint-Mard (Aisne), cant. Coucy-le-Château. 1296. III: Aisne, s.v.
Pont-Saint-Martin. Bridge on the Guiers-Vif between the communes Saint-Christophe-Entre-Deux-Guiers (Isère) and Saint-Christophe (Savoie). 15th cent. III: Isère, s.v.
Pont-Saint-Martin (Isère), communes Saint-Michel-les-Portes and Saint-Martin-de-Clelles. On the brook Saint Michel. 14th cent. III: Isère, s.v.
Pont-Saint-Martin (Loire-Inférieure), cant. Bouaze. 1179. III: Loire-Inférieure, s.v.
Pont-Saint-Ours, Le (Nièvre), com. Coulanges-lez-Nevers. 1251. III: Nièvre, s.v.
Pont-Saint-Pierre (Eure), arr. Les Andelys, cant. Fleury-sur-Andelle. 911–1066. III: Eure, s.v. IV: Adigard, 19.
Pont-Saint-Vincent (Meurthe), cant. Nancy-Ouest. 1177. III: Meurthe, s.v.
Pont-Sanguèze (Loire-Inférieure), com. Vallet. 1454. III: Loire-Inférieure, s.v.
Pontscorff (Morbihan), arr. Lorient, com. Cléguer. 1280. III: Morbihan, s.v.
Ponts-de-Cé, Les (Maine-et-Loire). Probably there in Caesar's time. Also 11th cent. IV: Bretaudeau, 39, 41.
Pont-Séchier (Isère), com. Livet-et-Gavet. On the Romanche. 14th cent. III: Isère, s.v.
Pont-Ségur (Isère), com. Mizoën. On the Romanche. 15th cent. III: Isère, s.v.
Pont-sur-Madon (Vosges), cant. Charmes. 1409. III: Vosges, s.v.
Pont-sur-Meuse (Meuse). 1106. III: Meuse, s.v.
Pont-sur-Seine (Aube), cant. Nogent-sur-Seine. Duodecim Pontes. 574, 864. Attested 1153. III: Aube, s.v.
Pont-sur-Vanne (Yonne), cant. Villeneuve-l'Archevêque. 1159. III: Yonne, s.v.
Pont-sur-Yonne (Yonne), arr. Sens. 833, 13th cent. III: Yonne, s.v.
Pont-Taulad (Aude), com. Cailhau. Name of a fief. 1166. III: Aude, *Dic.*, s.v.
Pont-Thibout, Le (Eure), hamlet of Trancheville. 1255. III: Eure, s.v.
Pont-Tranchêtu (Eure-et-Loir), communes Fontenay-sur-Eure and Nogent-sur-Eure. Ca. 1117. III: Eure-et-Loir, *Dic.*, s.v.
"Pontus" (Côte-d'Or), near Véronnes. 815. III: Côte-d'Or, s.v.
Pontvallain (Sarthe), arr. La Flèche. De ponte Valani. 1098. III: Sarthe, s.v.

Pontvianne (Haute-Loire), com. Solignac-sous-Roche. 1293. III: Haute-Loire, *Dic.*, s.v.
Pontvien, Le (Mayenne), com. Livré. Capellam de Pont-Viviani. 1184. III: Mayenne, *Dic top.*, s.v.
Pont-Vieux (Isère), com. Entraigues. Pons Veyl, on the Marsanne. 14th cent. III: Isère, s.v.
Pontvray (Marne), cant. Sillery. Pons Varensis, ca. 850. Pont-Veroit, 1227. III: Marne, s.v.
Pont-Yblon (Seine-et-Oise), hamlet Bonneuil, cant. Gonesse. 1093. II: Depoin, 1:76, #44.
Pont-Ysoir, Le (Sarthe), *see* Mans, Le.
Portejoie (Eure), cant. Pont-de-l'Arche, arr. Louviers. Turning bridge. 1198. II: Mortet and Deschamps, 172, n. 4.
Puy-de-Pont (Dordogne), com. Neuvic. 1203. III: Dordogne, s.v.
Radepont (Eure), cant. Fleury. Rotomagus in Antonine itinerary but Radipons in 1034. III: Eure, s.v.
Régennes, Les (Yonne), com. Appoigny. 1145. II: Quantin, *Cartulaire*, 2:394 ff.
Remi, *see* Sailly.
Rennepont (Haute-Marne), cant. Juzennecourt. Renepons. Ca. 1172. III: Haute-Marne, s.v.
Ribemont (Aisne), arr. St. Quentin. Pro pontibus de Ribermont. 1248. II: Mortet and Deschamps, 255–256.
Riom (Puy-de-Dome). Pont de Nonette. 1303. III: Riom, 74, CC58.
Ris, Le (Nièvre), com. Varennes-lez-Nevers. 1343. II: Bruel, 37, 44.
Robertière, La (Eure), near Saint-Georges-sur-Eure, arr. Evreux, cant. Nonancourt. 1346. II: Delisle, *Actes*, 336–342.
Rochelle, La (Charente-Maritime). 1202. II: Mortet and Deschamps, 194 ff.
Rochemaux (Vienne). On the Charente. Ca. 1080. IV: Crozet, 521.
Roche-Posay (Vienne). On the Creuse. 1175. IV: Crozet, 512.
Rochy-Condé (Oise), cant. Nivillers, arr. Beauvais. Pons de Craut prope villam de Condé. 1292. II: Leblond, 4:477.
Romans (Drôme). Bridge on the Isère between communes Romans and Bourg-de-Péage. From mid-11th cent. III: Drôme, *Dic.*, s.v.
Romorantin (Loire-et-Cher). 1336. III: Romorantin, 45 ff.
Roque, La (Gard), cant. Bagnols. On the Cèze. 13th cent. III: Gard, s.v.
Roquebrou, La (Cantal), arr. Aurillac. 1281–1282. II: Mortet and Deschamps, 302.
Roquecourbe (Tarn). On the Agout. Repaired 1317. IV: "Glanures," 7:240.
Rouanne (Oise), near Nanteuil-le-Haudouin. 13th cent. IV: Bourquelot, 316.
Rouen (Seine-Maritime). Early 11th cent. IV: Deville, 166 ff.
Sablé-sur-Sarthe (Sarthe). 13th cent. III: Sarthe, s.v.

Sailly (Somme), arr. Péronne, cant. Combles. Bridge between Remi and Sailly. 1236. II: Prarond, 162.
Saint-Affrique (Aveyron). On the Sorgue. Before 1368. IV: Emerson and Gromort, 64.
Saint-Antonin (Tarn-et-Garonne). On the Aveyron. 1358. II: Dumas, 279–280.
Saint-Astier (Dordogne). 1293. III: Dordogne, s.v.
Saint-Bertin (Pas-de-Calais), near Saint-Omer. On the Aa. 1177–1187. II: Mortet, *Recueil*, 122.
Saint-Cloud (Seine-et-Oise). 1411. II: Monstrelet, 203–206.
Saint-Denis (Seine). Pont Maubert, 1138. II: Suger, 346. Pont-la-Reyne, 1233. IV: Doublet, 3:905. "De ponte de Trecines" between Saint-Denis and Montque, rebuilt not long before 1247. IV: Champollion-Figeac, 132.
Sainte-Menehould (Marne). Pont de Berges, 1287. Pont de la Villeneuve, 1287. Pont de Gier, 1287. II: Longnon, 3:67.
Saintes (Charente-Maritime). Before 1201. IV: Viollet-le-Duc, 7:231. IV: Home, 343.
Saint-Flour (Cantal). Bridge on the Aude to reach suburb, before 1250. Two other bridges across the Aude to reach mills: Pont de Roueyre, 13th cent. Pont-du-Colombier, under construction in 1273. II: Boudet, ccxxviii, cccxi, 149.
Saint-Généroux (Deux-Sèvres). Traditionally 13th cent. IV: Emerson and Gromort, 49.
Saint-Geniez (Aveyron). 1356. II: *Ordonnances*, 3:158.
Saint-Geniez (Bouches-du-Rhône), at Martigues. Hospital of the bridge of Saint-Geniez. Before 1211. I: Bouches-du-Rhône, fols. 57v, 58v, 59r.
Saint-Guilhem-le-Désert (Hérault). Across the Gouffre-Noir of the Hérault. 1031–1048. II: Mortet, *Recueil*, 109 ff.
Saint-Jean-d'Angély (Charente-Maritime). A nearby bridge over the Charente was the scene of a fight in 1380. II: Froissart, 17:277.
Saint-Julien (Marne), com. Pierry. 1239. III: Marne, s.v.
Saint-Laurent-du-Pont (Isère), arr. Grenoble. 14th cent. IV: Sclafert, 212.
Saint-Lo (Manche). In Gallo-Roman times called Briovera. IV: Longnon, *Les Noms*, #98.
Saint-Maixent (Deux-Sèvres). Pont Charraut. 1366. II: Richard, 167.
Saint-Marceau (Sarthe), cant. Beaumont-sur-Sarthe, arr. Mamers. 12th cent. IV: Bienvenue, 218, 465.
Saint-Maur-de-Fossés (Seine). 1205. IV: Bousquié, 269.
Saint-Nicolas de Campagnac, *see* Campagnac.
Saint-Omer (Pas-de-Calais). "All the bridges over the river Aa between Saint-Omer and Gravelines." 1350. II: *Ordonnances*, 4:260–261.
Saint-Pierre-de-Maillé (Vienne). On the Gartemps. 1485. IV: Crozet, 511.
Saint-Pierre-du-Pont Neuf (Sarthe), com. and cant. Beaumont-sur-Sarthe, arr. Mamers. 12th cent. IV: Bienvenue, 435.

Saint-Pourçain (Allier). Under construction in 14th cent. IV: Langlois and Stein, 1:284.
Saint-Quentin (Aisne). 1237. II: Lemaire, 1:31.
Saint-Savin-sur-Gartemps (Vienne). Possibly 12th or 13th cent. IV: Crozet, 510.
Saint-Secondin (Vienne), cant. Gençay. 1329. IV: Crozet, 519.
Saint-Thibaut, *see* Sully-sur-Loire.
Salbris (Loire-et-Cher). Gallo-Roman. IV: Grenier, 2:264.
Salon (Bouches-du-Rhône). 1374. IV: Schäfer, 510.
Salonne (Meurthe). 950. II: Lauer, 79–80.
Saturargues (Hérault), cant. Lunel. Before 1140. II: Alaus, 1:478.
Saumur (Maine-et-Loire). Before 1162. II: Berger, 7,1:365.
Savines (Hautes-Alpes). 14th cent. IV: Sclafert, 710.
Séchilienne (Isère), com. and cant. Vizille. Ecclesia de S. Martini de Ponte-Roso. 10th cent. III: Isère, s.v.
Sens (Yonne). Roman, 519, 1145. IV: Quantin, 440.
Sérant (Loiret). Repairs 1387–1389. I: Loiret, A 1979, fol. L recto.
Serrières (Ardèche). 1251. IV: Guigue, *Cart. lyonnais*, 1:596.
Seyssel (between Ain and Haute-Savoie). Over the Rhône. A bridge there only from ca. 1300 to ca. 1322. IV: Châtelain, 118–119.
Sommières (Gard). On the Vidourle. Roman. Repaired in Middle Ages. IV: Emerson and Gromort, 31.
Sône, La (Isère). Bridge between communes La Sône and Saint-Just-de-Claix. 13th cent. III: Isère, s.v.
Sorgues (Vaucluse), arr. Avignon. 1125. IV: Devic and Vaissète, 4:75.
Sospel (Alpes-Maritimes). Possibly 11th cent. For description, see IV: Emerson and Gromort, 7.
Sully-sur-Loire (Loiret). Ancient, 1318. IV: Dion, 293 ff.
Suze, La (Sarthe), arr. Le Mans. 13th cent. IV: Bienvenue, 466.
Taillebourg (Cher-Maritime). "Cum pratis amoenis et ponte optimo." 1242. II: Paris, 589.
Tarascon (Bouches-du-Rhône). Bridge to château of counts of Anjou. 1448. II: Lecoy de la Marche, 138.
Tennie (Sarthe), cant. Conlie, arr. Mamers. Ca. 1216. IV: Bienvenue, 467.
Terrason (Dordogne). 1233. IV: Verneilh, 296.
Terride-en-Gimoës (Gers), arr. Lombers, cant. Cologne, com. St. Georges. 1295–1314. III: Toulouse, 1:134.
Thouars (Deux-Sèvres). 13th cent. IV: Emerson and Gromort, 11, 15, 47.
Thourotte (Oise), cant. Ribécourt. 13th cent. IV: Lot and Fawtier, CLXX.
Tocane (Dordogne). Pont de Pardutz. 1150. III: Dordogne, s.v.
Tonnerre (Yonne). Bridges. 1241. II: Quantin, *Recueil*, 215, #475.
Toulouse (Haute-Garonne). Old Bridge attested 1152. Pont-de-la-Daurade or New Bridge before 1181. IV: Wolff, 63. Pont-de-Bazacle, before 1219. II: Guillaume de Tudèle, line 9542. Bridges on the Hers: Pont de Pericole

(or de Negueromieu or de Saint Hilaire) and Pont de Velours (or Isalguier), 1282. I: Toulouse, DD 216. Pont de Las Clèdes, 1218. IV: Mundy, 105, n. 16.

Tournon (Ardèche). On the Doux. Begun 1351. IV: Séjourné, 2:16, 35 ff.

Tours (Indre-et-Loire). Over the Loire, 1031–1037. IV: Martène and Durand, 1:175–176. Over the Cher, 1172–1178. II: Berger, 7, 2:84.

Treix (Haute-Marne). Rebuilt 862. II: *Annales Bertiniani*, 57.

Trilbardou on the Marne, *see* Treix.

Trinité, La, or La Trinité-Victor (Alpes-Maritimes). Ancient abutments of bridge of Moulin d'Ezé. IV: Grenier, 2:732.

Troyes (Aube). Pont-Saint-Marie, 1st cent. Troyes, 1136. Pont-de-la-Salle, 1157. III: Aube, s.v. Le Pont-Hubert, com. Pont-Saint-Marie, 1154. IV: Boutiot, 1:209, 453. Pont-en-Bourbereau, 1374. IV: Boutiot, 1:51, n. Pont de Cailles over a canal, 13th cent. IV: Boutiot, 1:441. Le grand pont des Moulins-aux-Monts, com. Saint-Parres-aux-Tertres, and le pont Martinot, 1416. IV: Boutiot, 2:352. Pont de Sencey, 3rd canton of Troyes, 1433. IV: Boutiot, 2:578. Pont de Saint-Parres-les-Tertres, 1st canton of Troyes, and Pont de Fouchy, com. Lavan, 1431. IV: Boutiot, 2:529.

Uzès (Gard). Ca. 860. II: Poupardin, *Recueil Provence*, 13 ff.

Vaas (Sarthe), cant. Mayet, arr. La Flèche. Repairs 1382. IV: Bourmont, 173–174.

Vabres (Aveyron). On the Dourdou. 1302. III: Haute-Loire, *Dic.*, s.v. IV: Rey, 17, gives the date as 1277.

Valence (Drôme). 1214. IV: Mellier, 151–152.

Varennes-sur-Allier (Allier). 1258. II: Guigue, *Cartulaire lyonnais*, 2:58 ff.

Verdun (Meuse). 6th cent. II: Gregory of Tours, 3:26.

Verlhac-Tescou (Tarn-et-Garonne), arr. Montauban, cant. Villebrumieu. 1306. IV: Perbosc, 134.

Vermanton (Yonne). Over the Cure. 1238. IV: Quantin, 485.

Vernay, *see* Airvault.

Vernon (Eure). Before 1223. IV: Champollion-Figeac, 132.

Vicq-sur-Gartemps (Vienne). 1285. IV: Crozet, 510.

Vieille-Brioude (Haute-Loire). On the Allier. 1340. IV: Séjourné, 2:10.

Vieillevie (Cantal). 1218. II: Desjardins, 391.

Vienne (Isère). Replacing old bridge. 1251. II: Guigue, *Cartulaire lyonnais*, 1:576.

Villefranche-de-Belvez (Dordogne), arr. Sarlat. Formerly Villefranche-en-Périgord. 1357. II: *Ordonnances*, 3:206.

Villemanoche (Yonne), cant. Pont-sur-Yonne. 823. II: Quantin, 1:41.

Villeneuve-sur-Lot (Lot-et-Garonne). 1282. IV: Samazeuilh, 1:326.

Villeneuve-sur-Yonne (Yonne), arr. Joigny. Formerly Villeneuve-le-Roi. 1186. IV: Quantin, 439.

Villerets (Eure), com. Ecouais. 1260. II: *Olim*, 1:496.

Vitry-le-François (Marne). Two bridges, one to a castle. 1267.
II: Longnon, 3:67, 68.
Vivonne (Vienne). Legacy for repairs to bridge. 1264. IV: Crozet, 516.
Vizille (Isère). Washed out in 1336. IV: Sclafert, 171.
Warnéton (Nord), cant. Quenoy-sur-Deule, arr. Lille. 1093–1111.
II: Depoin, 1:#138.
Youle (Aveyron). On the Aveyron. 1320. IV: Rey, 17.

Abbreviations

Acad. Besançon: Académie des Sciences, Belles-lettres et Arts de Besançon.
Annales arch.: Annales archéologiques.
Annales Avignon: Annales d'Avignon et du Comtat Venaissin.
Annales Gâtinais: Annales de la Société historique et archéologique de Gâtinais.
Annales Mâcon: Annales de l'Académie de Mâcon, Société des Arts, Sciences, Belles-lettres, Archéologie, Agriculture et Encouragement au Bien du Saône-et-Loire.
Annuaire Lorraine: Annuaire de la Société d'histoire et d'archéologie de la Lorraine.
Arch. Gironde: Archives historiques de la Gironde.
Arch. Maine: Archives historiques de Maine.
Arch. Poitou: Archives historiques du Poitou.
Arch. Rouergue: Archives historiques du Rouergue.
Arch. Saintonge: Archives historiques de la Saintonge et de l'Aunis.
Bibl. Ecole Chartes: Bibliothèque de l'Ecole de Chartes.
Bibl. Ecole franç. Rome: Bibliothèque des Ecoles français d'Athènes et de Rome.
Bibl. Ecole Haute-Etudes: Bibliothèque de l'Ecole des Hautes-Etudes. Sciences philologiques et historiques.
Bull. Drôme: Bulletin de la Société départementale d'archéologie et de statistique de la Drôme.
Bull. Mayenne: Bulletin de la Commission historique et archéologique de la Mayenne.
Bull. Midi: Bulletin de la Société archéologique du Midi de la France.
Bull. Orléanais: Bulletin de la Société archéologique et historique de l'Orléanais.
Bull. Ouest: Bulletin de la Société des Antiquaires de l'Ouest.
Bull. Paris: Bulletin de la Société de l'Histoire de Paris.
Bull. Périgord: Bulletin de la Société historique et archéologique du Périgord.
Bull. Sens: Bulletin de la Société archéologique de Sens.
Bull. Tarne-et-Garonne: Bulletin de la Société archéologique de Tarn-et-Garonne.

Bull. Touraine: Bulletin de la Société archéologique de Touraine.

Bull. Vaucluse: Bulletin historique et archéologique de Vaucluse.

Bull. Yonne: Bulletin de la Société des Sciences historiques et naturelles de l'Yonne.

Chartes Hist. France: Chartes et diplômes relatifs à l'histoire de France au Moyen-Age.

Classiques: Classiques de l'histoire de France au Moyen Age.

Collection de documents Monaco: Collection de documents historiques, pub. par ordre de S.A.S. le prince Albert Ier, prince souverain de Monaco.

Collection de textes histoire: Collection de textes pour servir à l'étude et à l'enseignement de l'histoire.

Collection de textes de Provence: Collection de textes pour servir à l'histoire de Provence.

Comité des Travuax hist.: Comité des Travaux historiques et scientifiques, Bulletin philologique et historique.

Ecole pratique. VIe sec.: Ecole pratique des Hautes-Etudes. VIe section. Centre de recherches historiques. Affaires et gens d'affaires.

Historiens de France: Recueil des historiens des Gaules et de la France.

Inst. textes: Institut de Recherches et d'Histoire des Textes. Editions du Centre national de la Recherche Scientifique.

Inv. et doc.: Archives de l'Empire. Inventaires et documents, pub. par ordre de l'empéreur sous la direction de M. le comte de Laborde.

Layettes: Layettes du trésor des chartes.

Mélanges Ecole franç. de Rome: Mélanges d'Archéologie et d'Histoire de l'Ecole française de Rome.

Mém. Abbeville: Mémoires de la Société d'Emulation d'Abbeville.

Mém. Acad. Gard: Mémoires de l'Académie du Gard.

Mém. Acad. Inscriptions: Mémoires de l'Académie des Inscriptions et Belles-lettres.

Mém. Acad. Metz: Mémoires de l'Académie impériale de Metz.

Mém. Acad. Nîmes: Mémoires de l'Académie de Nîmes.

Mém. Acad. Toulouse: Mémoires de l'Académie des Sciences, Inscriptions, et Belles-Lettres de Toulouse.

Mém. Acad. Vaucluse: Mémoires de l'Académie de Vaucluse.

Mém. Angers: Mémoires de la Société nationale d'agriculture, Sciences et Arts d'Angers.

Mém. Féd. Paris: Mémoires de la Fédération des Sociétés historiques et archéologiques de Paris et de l'Ile-de-France.

Mém. Loire-et-Cher: Mémoires de la Société des Sciences et lettres de Loire-et-Cher.

Mém. Lyon: Mémoires de la Société littéraire de Lyon.

Mém. Midi: Mémoires de la Société archéologique du Midi de la France.

Mém. Nivernais: Mémoires de la Société académique de Nivernais.

Mém. Orléanais: Mémoires de la Société archéologique et historique de l'Orléanais.

Mém. Paris: Mémoires de la Société de l'Histoire de Paris.
Mém. Savoisienne: Mémoires et documents publiées par la Société Savoisienne d'Histoire et d'Archéologie.
Mém. Touraine: Mémoires de la Société archéologique de Touraine.
MGH AA: Monumenta Germaniae Historica. Auctores antiquissimi.
MGH Capit.: Monumenta Germaniae Historica. Capitularia regum Francorum.
MGH SSrerMerov: Monumenta Germaniae Historica. Scriptores rerum Merovingicarum.
MGH SSrerG: Monumenta Germaniae Historica. Scriptores rerum Germanicarum.
Olim: Les Olim ou registres de arrêts rendus par la cour du roi sous les règnes de Saint Louis, de Philippe le Hardi, de Philippe le Bel, de Louis le Hutin et de Philippe le Long.
Ord.: Ordonnances des roys de France de la troisième race, recueillies par ordre chronologiques.
Rev. arch.: Revue archéologique.
Rev. géog. Pyrénées: Revue géographique des Pyrénées et du Sud-Ouest.
Rev. Maine: Revue historique et archéologique du Maine.
Rev. Tarn: Revue historique, scientifique et littéraire du département de Tarn.
Rev. Soc. Savantes: Revue des Sociétés savantes.
Soc. Acad. Saint-Quentin: Société académique des Sciences, Arts, Belles-Lettres, Agriculture et Industrie de Saint-Quentin.
Soc. Agen: Société d'agriculture, sciences et arts d'Agen. Archives historiques de l'Agenais.
Soc. Angers: Société d'Agriculture, Sciences et Arts d'Angers. Documents historiques sur l'Anjou.
Soc. Auvergne: Société d'Emulation d'Auvergne.
Soc. Compiègne: Société historique de Compiègne.
Soc. Dordogne: Société d'Agriculture, Sciences et Arts de Dordogne.
Soc. Ecole Chartes: Société de l'Ecole des Chartes.
Soc. Eure: Société libre d'Agriculture, Science, Arts et Belles-lettres de l'Eure.
Soc. Eure-et-Loir: Société archéologique d'Eure-et-Loir.
Soc. France: Société de l'Histoire de France.
Soc. Gascogne: Société historique de Gascogne.
Soc. Lyon: Société littéraire, historique et archéologique de Lyon.
Soc. Montpellier: Société archéologique de Montpellier.
Soc. Nivernaise: Société Nivernaise des Lettres, Sciences et Arts.
Soc. Normandie: Société de l'Histoire de Normandie.
Soc. Oise: Société académique de l'Oise.
Soc. Paris: Société de l'Histoire de Paris.
Soc. Seine-et-Marne: Société d'Archéologie, Sciences, Lettres et Arts du Département de Seine-et-Marne.
Soc. Touraine: Société archéologique de Touraine.
Soc. Yonne: Société des Sciences historiques et naturelles de l'Yonne.

Travaux Acad. Reims: Travaux de l'Académie de Reims.
Travaux Acad. Rouen: Travaux de l'Académie royale des Sciences, Belles-lettres et Arts de Rouen.

Bibliography

I. Manuscript Sources

Albi. Archives communales. CC 167, 168, 180.

Bouches-du-Rhône. Archives départementales. Livre Trésor de l'hôpital Saint-Esprit de Marseille, 1399, fols. 57 verso, 58 verso, 59 recto.

Haute-Garonne. Archives départementales. H. Daurade 8: B 1, fols. 7, 225, 244, 355; B 21, fols. 163, 169; B 3, fol. 21; B 4, fols. 264, 309, 369; B 6, fol. 21.

Loiret. Archives départementales. Archives hospitalières III A 1. A 1979, fols. 49, 50.

Orleans. Archives communales. CC 920, 930, 967.

Paris. Bibliothèque Nationale. Collection Doat 60, fols. 123 recto and verso, 124 recto; Doat 93, fols. 210 verso, 211 recto; Doat 118, fol. 116 verso; Doat 127, fol. 145 recto and verso; Doat 130, fols. 259 recto, 298 verso; Doat 131, fol. 291 recto and verso.

Toulouse. Archives communales. DD 154 (anc. DD5); DD 211 (anc. 242); DD 216 (anc. 249).

Vaucluse. Archives départementales. G 5 no 38. Fonds de Bonpas, H. 1189, 1197, 1212, 1277, 1278, 1284.

II. Published Sources

Abbo. *Le Siège de Paris par les Normandes. Poème du IXe siècle.* Ed. and trans. Henri Waquet. Les Classiques, 20. Paris, 1942.

Acta Sanctorum. April, vol. 2. New ed. by John Carnandet. Paris and Rome, 1865.

Ado. "Ex Adonis archiepiscopi Viennensis chronico." *Historiens de France*, 7:54–56.

Alaus, Paul, Abbé Casson, and Meynial, eds. *Cartulaires des abbayes d'Aniane et de Gellone publié d'après les manuscrits originaux.* 2 vols. Soc. Montpellier. Montpellier, 1898–1900.

Alberti, Leone Battista. *Ten Books on Architecture.* English trans. by James Leoni, first pub. 1726. Ed. Joseph Rykwert. New York, 1966.

Annales Bertiniani. Ed. G. Waitz. MGH SSrG. Hanover, 1883.

Annales Xantenses et Annales Vedastini. Ed. B. de Simson. MGH SSrG. Hanover and Leipzig, 1909.

Arleta, Antionio Urbieto, ed. *Cartulario de San Juan de la Peña.* Textos Medievales, 9. Valencia, 1963.

Artières, Jules, ed. *Documents sur la ville de Millau. Mémorial des Privilèges. Livre de comptes des Consuls Boursiers. Délibérations (XIe–XVIe siècles).* Arch. Rouergue. Millau, 1930.

Basin, Thomas. *Histoire de Charles VII.* Ed. and trans. Charles Samaran. Les Classiques, 15. Paris, 1933.

Beaumanoir, Philippe de. *Coutumes de Beauvaisis.* Ed. Am. Salmon. Collection de textes histoire. 2 vols. Paris, 1899–1900.

Bémont, Charles, ed. *Rôles Gascons.* Collection de documents inédits, 2, 3. Paris, 1900–1906.

Berger, Elie, ed. *Recueil des actes de Henri II, roi d'Angleterre et duc de Normandie, concernant les provinces françaises et les affaires de France.* Chartes Hist. France, 4, 7. Paris, 1909–1927.

Bernardus Guidonis. "E Floribus chronicum per catalogo Romanorum pontificium." *Historiens de France*, 21:706.

Bernardus Iterii. "Ex Chronico Bernardi Iterii monachi et armarii Sancti Martialis Lemovicensis." *Historiens de France*, 18:234.

Blanc, Alphonse, ed. *Le Livre de comptes de Jacme Olivier, marchand Narbonnais du XIVe siècle.* Vol. 2, part 1. No more pub. Paris, 1899.

Boileau, Etienne. *Les Métiers et Corporations de la Ville de Paris. XIIIe siècle. Le Livre des Métiers d'Etienne Boileau.* Ed. René de Lespinasse and François Bonnardot. Histoire générale de Paris. Paris, 1879.

Bonnin, Théodore, ed. *Cartulaire de Louviers: documents historiques originaux du Xe au XVIIIe siècle, la plupart inédits.* 5 vols. Evreux, 1879–1883.

Boudet, Marcellin, ed. *Cartulaire du prieuré de Saint-Flour.* Preface by A. Bruel. Collection de documents Monaco. Monaco, 1910.

Bourassé, Jean Jacques, ed. *Cartulaire de Cormery précédé de l'histoire de l'abbaye et de la ville de Cormery d'après chartes.* Paris, 1861.

Bourquelot, Félix. "Fragments de comptes du XIIIe siècle." *Bibl. Ecole Chartes* 24 (1863), 51–79.

Boutaric, E., ed. *Actes du parlement de Paris.* 2 vols. Inv. et doc. Paris, 1863–1867.

Breuils, L'Abbé, and Gardère, eds. "Comptes des consuls de Montréal-du-Gers." *Arch. Gironde* (1894), pp. 283–328.

Brièle, Léon, and Ernest Coyecque, eds. *Archives de l'Hôtel-Dieu de Paris (1157–1300).* Collection de documents inédits. Paris, 1894.

Bruel, Alexandre. *Visites des monastères de l'ordre de Cluny de la province d'Auvergne aux XIIIe et XIVe siècles. Nouvelle Série.* Paris, 1891.

Bruguier-Roure, L., ed. *Cartulaire de l'oeuvre des église, maison, pont et hôpitaux du Saint-Esprit (1265–1791).* Mém. Acad. Nîmes, ser. 7, Annexes to vols. 12 (1889), 13 (1890), 14 (1891), 16 (1893), 17 (1894).

Caesar, C. Julius. *De bello gallico.* Ed. St. George William Joseph Stock. Oxford, 1898.

Capitularia regum Francorum. Ed. Alfred Boretius and Victor Kraus. MGH Capit. Hanover, 1897.

Catel, Albert, and Maurice Lecomte, eds. *Chartes et documents de l'abbaye cistercienne de Preuilly.* Soc. Seine-et-Marne. Paris, 1927.

Chantelou, Claude. *Cartulaire tourangeau et sceaux des abbés.* [Marmoutier, France] Ed. Paul Nobilleau. Tours, 1879.

Charmasse, A. de, ed. *Cartulaire de l'église d'Autun.* 3 parts. Société Eduenne. Paris and Autun, 1865–1900.

Charpin-Feugerolles, Le Comte de, and Marie-Claude Guigue, eds. *Grand Cartulaire de l'abbaye d'Ainay, suivi d'un autre cartulaire redigé en 1286.* 2 vols. Lyons, 1885.

Chastellain, Georges. *Oeuvres.* Ed. Kervyn de Lettenhove. 8 vols. Brussels, 1863–1866.

Chevalier, C.U.J., ed. *Choix de documents historiques inédits sur le Dauphiné.* Montbéliard and Lyons, 1874.

Chrétien de Troyes. *Arthurian Romances.* Trans. W. Wistar Comfort. New York, 1914, 1928.

Clement V. *Regestum Clementis papae V.* Ed. monks of order of St. Benedict. 9 vols. Rome, 1884–1888.

Coville, Alfred, ed. *L'Ordonnance cabochienne (26–27 mai 1413).* Collection de textes histoire. Paris, 1891.

De Broussillon, Bertrand, ed. *Cartulaire de l'évêché du Mans (936–1790).* Arch. Maine, 9. Le Mans, 1900.

Delaborde, H.F., ed. *Recueil des actes de Philippe Auguste.* Chartes. Hist. France. Paris, 1916.

Delachenel, R., ed. *Les Grandes Chronques de France. Chroniques des règnes de Jean II et de Charles V.* Soc. France. 4 vols. Paris, 1910–1920.

Delaville-Le-Roulx, Joseph M.A., ed. *Registres de comptes municipaux de la ville de Tours.* Soc. Touraine. 2 vols. Tours, 1878–1881.

Delisle, Léopold, ed. *Actes Normandes de la chambre des comptes sous Philippe de Valois (1328–1350).* Soc. Normandie. Rouen, 1871.

_____. *Mandements et actes divers de Charles V (1364–1380) recueillis dans les collections de la Bibliothèque nationale.* Collection de documents inédits. Paris, 1874.

Delpit, Jules. "Vidimus par l'officiel de Bordeaux d'un jugement du sénéchal pour le péage des Pèlerins." *Arch. Gironde* 6 (1864), 1–2.

"De origine comitum andegavensium." *Historiens de France*, 12:535–536.

Depoin, J., ed. *Recueil de chartes et documents de Saint-Martin-des-Champs, monastère parisien.* 5 vols. Archives de la France monastique, 13, 16, 18, 20, 21. Paris and Chevetogne, Belgium, 1912–1921.

De Richemond, L.M. "Chartes de la commanderie magistrale du Temple de la Rochelle (1139–1268)." *Arch. Saintonge* 1 (1874), 21–50.

Desjardins, Gustave, ed. *Cartulaire de Conques.* Soc. Ecole Chartes. Paris, 1879.

Devaux, A. *Comptes consulaires en langue vulgaire (1338–1340).* Montpellier, 1912.

De Wailly and Delisle, eds. "Compotus praepositorum et baillivorum Franciae de termino ascensionis, anno domini MCCXLVIII." *Historiens de France*, 21:260–284.

Drouyn, Leo, ed. *Comptes de l'archevêché de Bordeaux du XIIIe et du XIVe siècle. Arch. Gironde*, 21. Bordeaux, 1881.

Druilhet, P., ed. *Archives de la ville de Lectoure, coutumes, statuts et records du XIIIme au XVme siècle.* Soc. Gascogne. Paris, 1885.

Du Brossay, M. "Cartulaire d'Azé et du Géneteil, prieuré de l'abbaye Saint-Nicolas d'Angers." *Arch. Maine* 3 (1903), 49–168.

Du Cange, C.D. *Glossarium mediae et infimae latinitatis.* 10 vols. Reissued by L. Favre. Paris, 1937–1938.

Ducéré, Edouard, and Pierre Yturbide, eds. *Livre des établissements. Archives municipales de Bayonne.* Bayonne, 1892.

Dumas de Rauly, Charles, "Documents inédits sur Saint-Antonin pendant la guerre de cent ans. Extraits de l'inventaire sommaire des archives de cette ville." *Bull. Tarn-et-Garonne* 9 (1881), 273–301.

Einhard. *The Life of Charlemagne.* Trans. Samuel E. Turner. First pub. 1880. Ann Arbor, 1960.

"Ex chronico S. Albini Andegavensis (ab. an. DCCCCXXIX ad an. MCC)." *Historiens de France*, 12:481.

"Ex gestis ambasiensium dominorum." *Historiens de France*, 12:510.

"Ex gestis consulum andegavensium. Auctore monacho Benedictino Majoris monsaterio." *Historiens de France*, 10:255.

"Ex gestis pontificium Cenomanesium." *Historiens de France*, 12:554.

"Ex miraculis sanctae Genovefae." *Historiens de France*, 18:797–798.

Flower, Cyril Thomas, ed. *Public Works in Medieval Law.* 2 vols. Selden Society. London, 1915–1923.

Fortunatus. "Vita sancti Leobini." Ed. Bruno Krusch. MGH AA 4,2:79.

"Fragmentum genealogicum ducum Normanniae et Angliae regum." *Historiens de France*, 18:242.

Fredegarius Scholasticus. *Chronicarum quae dicuntur Fredegarii Scholastici Libri IV cum continuationibus.* Ed. Bruno Krusch. MGH SSrerMerov, 2. Hanover, 1888.

Froissart, Jean. *Oeuvres.* Ed. Kervyn de Lettenhove. 25 vols. in 26. Brussels, 1870–1877.

Godfrey, canon of St. Victor. "Fons philosophiae." *Historiens de France*, 18:798.

Grandmaison, Louis de, ed. *Cartulaire de l'archevêché de Tours* (*Liber bonarum gentium*). 2 vols. Soc. Touraine, 37–38. Tours, 1892–1894.

Gregory the Great. *Dialogi Libri IV*. Ed. Umberto Moricca. Fonti per la storia d'Italia, 57. Rome, 1924.

Gregory IX. *Les Registres. Recueil des bulles de ce pape*. Ed. L. Auvray. 4 vols. Paris, 1896–1955.

Gregory of Tours. *Historia Francorum*. Ed. William Arndt. MGH SSrerMerov, 1. Hanover, 1885.

_____. *History of the Franks*. Trans. Ernest Brehaut. Selections trans. with notes. Records of Civilization, Sources and Studies, 11. New York, 1916.

_____. *History of the Franks*. Trans. with an introduction by O.M. Dalton. Oxford, 1927.

Grosse-Duperon, A., and E. Gouvrion, eds. and transs. *Cartulaire de l'abbaye cistercienne de Fontaine-Daniel*. Mayenne, 1896.

Guérard, Benjamin E.C., ed. *Cartulaire de l'église de Notre-Dame de Paris*. 4 vols. Collection de documents inédits. Paris, 1850.

Guérin, Paul, ed. *Recueil des documents concernant le Poitou contenus dans les registres de la chancellerie de France*. 14 vols. Arch. Poitou. Poitiers, 1881–1958.

Guigue, Marie Claude, ed. *Cartulaire lyonnais; documents inédits pour servir à l'histoire des anciennces provinces de Lyonnais, Forez, Beaujolais, Dombes, Bresse et Bugez, compris jadis dans le Pagus major Lugdunensis. . . .* 2 vols. Lyons, 1885–1893.

_____, ed. *Cartulaire municipal de la ville de Lyon: privilèges, franchises, libertés et aûtres titres de la commune; recueil formé au XIVe siècle par Etienne de Villeneuve*. Soc. Lyon. Lyons, 1876.

_____. *Obituarium Lugdunensis ecclesiae. Nécrologie des personnages illustres et des bienfaiteurs de l'Eglise métropolitaine de Lyon du IXe au XVe siècle*. Lyons, 1867.

_____, ed. *Registres consulaires de la ville de Lyon ou Recuil des déliberations du conseil de la commune de 1416 à 1423*. Lyons, 1882.

Guillaume de Tudèle. *La Chanson de la Croisade contre les Albigeois commencée par Guillaume de Tudèle et continuée par un poète anonyme*. Ed. and trans. Paul Meyer. 2 vols. Soc. Paris. Paris, 1875–1879.

Halphen, Louis, ed. *Receueil d'annales angevines et Vendomoises*. Collection de textes histoire. Paris, 1903.

_____, and René Poupardin, eds. *Chroniques des comtes d'Anjou et des seigneurs d'Amboise*. Collection de textes histoire. Paris, 1913.

Higounet, Charles, "Un Compte de ferme de la barre d'Agen au debut du XIVe siècle." *Annales du Midi* 62 (1950), 351–355.

Homer. *Iliad*. Trans. A.T. Murray. 2 vols. Loeb Classical Library. Cambridge, Mass. 1965–1967.

Isambert, Jourdan, and Decrusy, eds. *Receuil général des anciennes lois françaises depuis l'an 420 jusqu'à la révolution de 1789*. 29 vols. Paris, 1822–1833.

Izarn, E., ed. *Le Compte des recettes et dépenses du roi de Navarre en France et en Normandie de 1367 à 1390*. Introd. Gustave A. Prévost. Paris, 1885.

Jeanton, G. "Documents pour servir à l'histoire de Bourgogne. Comptes de la châtellenie de Cuisery au XIVe siècle." *Annales Mâcon*, 3rd ser., 19 (1914–1915), 93–191.

John XXII. *Lettres communales analysées d'après les registres dits d'Avignon et du Vatican.* Ed. G. Mollat. Bibl. Ecole franç. Rome, 1. Paris, 1904.

Joinville, Jean, Sire de. *Histoire de Saint Louis.* Original text. Ed. Natalis de Wailly. N.P., 1874.

Julian. *Works of Emperor Julian.* Trans. Wilmer Cave Wright. Loeb Classical Library. London and New York, 1913.

Lasteyrie du Saillant, Robert Charles, Count of, ed. *Cartulaire général de Paris, ou Recueil de documents relatifs à l'histoire et topographie de Paris.* Paris, 1887.

La Trémoille, Louis, Duke of, ed. *Chartrier de Thouars; documents historiques et généaologiques.* Paris, 1877.

Lauer, Philippe, ed. *Recueil des actes de Louis IV, roi de France (936–954).* Chartes hist. France. Paris, 1914.

Layettes du trésor des chartes. Ed. Alexandre Teulet. 2 vols. Inv. et doc. Paris, 1863–1866.

Leblond, V., ed. *Cartulaire de l'hôtel-dieu de Beauvais compremant 529 chartes la plupart originales conservées aux archives hospitalières de cette ville.* Soc. Oise, 4. Paris, 1919.

Lecoanet, Simone. "Charte de Eudes II, comte de Blois, concernant le premier pont de Tours." *Bull. Touraine* 35 (1968), 244–247.

Lecoy de la Marche, A., ed. *Extraits des comptes et mémoriaux du roi René, pour servir à l'histoire des arts au XVe siècle; publiés d'après les originaux des archives nationales.* Paris, 1873.

Lemaire, Emmanuel, ed. *Archives anciennes de la ville de Saint-Quentin. Précédées d'une etude sur les origines de Saint-Quentin*, by A. Giry. 2 vols. Soc. Acad. Saint-Quentin. Saint-Quentin, 1888.

Le Roux de Lincy, A.J.V., and L.M. Tisserand, eds. *Paris et ses historiens au XIVe et XVe siècles.* Histoire Générale de Paris. Paris, 1867.

Livy. *Periochae or Summaries.* In *Works*, 14, ed. B.O. Forster. Loeb Classical Library. Cambridge and London, 1959.

Longnon, Auguste, ed. *Documents relatifs au comté de Champagne et de Brie. 1172–1361.* 3 vols. Collection de documents inédits. Paris, 1901–1914.

Magen, Adolphe, ed. and trans. *Jurades de la ville d'Agen (1345–1355).* Soc. Agen. Auch, 1894.

Mahul, Alphonse Jacques, ed. *Cartulaire et archives des communes de l'ancien diocèse et de l'arrondissement administratif de Carcassonne.* 6 vols. in 7. Paris, 1857–1885.

Marchegay, Paul, ed. *Archives d'Anjou: receuil de documents et mémoires inédits sur cette province.* 3 vols. in 2. Angers, 1843–1864.

_____. "Chartes angevines des onzième et douzième siècles." *Bibl. Ecole Chartes* 35 (1875), 391–456.

———, and Emile Mabille, eds. *Chronique des églises d'Anjou.* Soc. France. Paris, 1899.

Martène, Edmund, and Ursin Durand, eds. *Thesaurus novus anecdotorum.* 5 vols. Paris, 1717.

Michon, [L.], "Documents relatifs à l'histoire de la ville de Mâcon (1362–1367)." *Rev. Soc. Savantes,* 4th ser., 9 (1869), 474–475; 5th ser., 1 (1870), 182–183.

Monstrelet, Enguerrand de. *Chronique.* Ed. L. Douët-D'Arcq. 6 vols. Soc. Paris. Paris, 1857–1862.

Morel, Emile Epiphanius, ed. *Cartulaire de Saint-Corneille de Compiègne.* 2 vols. Soc. Compiègne. Montdidier, 1909.

Mortet, Victor. "Un Formulaire du VIIIe siècle pour les fondations d'édifices et de ponts d'après des sources d'origine antique. Nouvelle édition critique." *Bulletin monumentale* 71 (1907), 442–465.

———, ed. *Recueil de textes relatifs à l'histoire de l'architecture et à la condition des architectes en France, au moyen âge, XIe–XIIe siècles.* Collection de textes histoire, 44. Paris, 1911.

———, and Paul Deschamps, eds. *Recueil de textes relatifs à l'histoire de l'architecture et à la condition des architectes en France, XIIe–XIIIe siècles.* Collection de textes histoire, 51. Paris, 1929.

Mugnier, F. "Comptes de la châtellenie de la Balme en Genevois." *Mém. Savoisienne* 30 (1891), 377–480.

Nanglard, J., ed. *Cartulaire de l'église d'Angoulême.* Angoulême, 1900.

Olim, Les, ou registres de arrêts rendus par la cour du roi sous les règnes de Saint Louis, de Philippe le Hardi, de Philippe le Bel, de Louis le Hutin et de Philippe le Long. Ed. A.A. Beugnot. 3 vols. in 4. Paris, 1839–1848.

Ordonnances des roys de France de la troisième race, recuillies par ordre chronologique. 22 vols. Paris, 1823–1849.

Orosius, Paul. *Seven Books against the Pagans: The Apology.* Trans. I.W. Raymond. Records of Civilization. Sources and Studies, 26. New York, 1936.

Paris, Matthew. *Chronica majora.* Ed. Henry Richards Luard. 7 vols. Rerum britannicarum medii aevi, 57. London, 1872–1883.

Perbosc, Antonin, and Severin Canal. "Chartes de coutumes du Quercy en Langue d'oc. Coutumes de Verlhac-Tescou." *Bull. Tarn-et-Garonne* 55 (1927), 120–141.

Phillipps, Thomas. "Letter . . . communicating a transcript of a MS Treatise on the preparation of Pigments, and on various processes of the Decorative Arts practiced in the Middle Ages, written in the twelfth century, and entitled Mappae clavicula," *Archaeologia* 32 (1847), 183–244.

Planchenault, Adrien, ed. *Cartulaire du chapitre de Saint-Laud d'Angers (actes du XIe et du XIIe siècle) suivi de la vie de Saint Silvestre et l'invention de la Sainte Croix. Poème français du XIIe siècle.* Soc. Angers, 4. Angers, 1903.

Pliny the elder. *Naturalis Historiae libri XXXVII.* Trans. H. Rackham, W.H.S. Jones, D.E. Eichholz. 10 vols. Loeb Classical Library. Cambridge, Mass., 1938–1962.

Porter, Arthur Kingsley. *Lombard Architecture.* 4 vols. New Haven and London, 1917.

Poupardin, René, ed. *Recueil des actes des rois de Provence (855–928).* Chartes Hist. France, 5. Paris, 1920.

_____. *Recueil des chartes de l'abbaye de Saint-Germain-des-Près des origines au début du XIIIe siècle.* 2 vols. 2nd revised by A. Vidier and Léon Levillain. Soc. Paris. Paris, 1909 and n. d.

Prarond, Ernest, ed. *Cartulaire du comte de Ponthieu.* Mém. Abbeville, 2. Abbeville, 1897.

Prinet, Max. "Document relatif à la chute du Pont Notre-Dame (1499)." *Mem. Paris* 49 (1927), 103–107.

Prou, Maurice. *Recueil des actes de Philippe I, roi de France (1059–1108).* Chartes Hist. France, 1. Paris, 1908.

Quantin, Maximilien, ed. *Cartulaire général de l'Yonne; receuil de documents authentiques pour servir à l'histoire des pays qui forment ce département.* 2 vols. Soc. Yonne. Auxerre, 1854–1860.

_____, ed. *Recueil de pièces pour faire suite au Cartulaire général de l'Yonne (XIIIe siècle).* Auxerre, 1873.

Ribere, Joseph Ignace de, chevalier seigneur de Costebelle, Esprit Roubert, Antoine Vincent et al., eds. *Statuts de la cité d'Avignon, avec la convention d'icelle, latin et françois, respondant par deux colomnes, l'une à l'autre, pour ceux qui veulent avoir l'intelligence de l'une d'icelles.* First printed 1613. Reprinted 1698.

Richard, Alfred. *Chartes et documents pour servir à l'histoire de l'abbaye de Saint-Maixent (suite).* Arch. Poitou, 18. Poitiers, 1886.

Richer. *Historiae Francorum.* Ed. and trans. Robert Latouche. Classiques, 12, 17. Paris, 1930–1937.

Rigord. "De gestis Philippi Augusti." *Historiens de France*, 17:61.

Ripert-Monclar, Francois, marquis de, ed. *Bullaire des indulgences concédées avant 1431 à l'oeuvre du pont d'Avignon par les souverains pontifes.* Collection de textes de Provence, 1. Monaco, 1912.

Robert of Auxerre, "Ex chronologia sancti Mariani altissiodorensis." *Historiens de France*, 12:298.

Robert of Torigny. "Appendice ad Sigebertum ad calcem operum Guiberti Novigenti abbatis." *Historiens de France*, 13:290.

Saige, Gustave, and Henri Lacaille. *Trésor des chartes du comté de Rethel.* 4 vols., 3 and 4 ed. L.H. Labande. Collection de documents Monaco. Monaco, 1902–1916.

Schäfer, K.H. *Die Ausgaben der apostolischen Kammer unter den Päpsten Urban V and Gregor XI (1362–1378).* Vatikanische Quellen für Geschichte des Päpstlichen Hof- und Finanzverwaltung. Görresgesellschaft, 6. Paderborn, 1927.

Sidonius Apollinarius. *Poems and Letters.* Trans. W.B. Anderson. Loeb Classical Library. Cambridge, Mass. 1936.

Smith, Lucy Toulmin, ed. *Expeditions to Prussia and the Holy Land made by Henry Earl of Derby (afterwards King Henry IV) in the years 1390–1 and 1392–3 being the accounts kept by his treasurer during those two years.* Camden Society. London, 1894.

Soyer, Jacques, and Guy Trouillard, eds. *Cartulaire de la ville de Blois (1196–1493); recueil manuscrit du XVe siècle conservé à la Bibliothèque Nationale.* Mém. Loire-et-Cher. Paris, 1907.

Strabo. *The Geography.* Trans. Horace Leonard Jones. 8 vols. Loeb Classical Library. Cambridge and London, 1948.

Suger. *Oeuvres complètes.* Ed. A. Lecoy de la Marche. Soc. France. Paris, 1867.

Tardif, Jules. *Monuments historiques [Cartons des rois], 528–1789.* Inv. et doc. Paris, 1866.

Terroine, A., and L. Fossier, eds., with Y. de Montenon. *Chartes et documents de l'abbaye de Saint-Magloire.* Inst. textes, 12, Chartriers des ancience abbayes de la ville de Paris, 2. Paris, 1966.

Theodosius. *Theodosiani libri XVI cum Consitutionibus Sirmondianis et Leges novellae ad Theodosianum pertinentes.* Ed. Th. Mommsen and Paul M. Meyer. Academia litterarum regiae borussica. Berlin, 1905.

Tholin, G. "Chartes d'Agen se rapportant au règne de Jean le Bon et de Charles V." *Arch. Gironde* 34 (1899), 147–210.

———. "Chartes d'Agen se rapportant au règne de Philippe de Valois." *Arch. Gironde* 33 (1898), 75–172.

Tuetey, Alexandre, ed. *Journal d'un bourgeois de Paris, 1405–1449.* Soc. Paris. Paris, 1881.

Urseau, Charles, ed. *Cartulaire noir de la cathédrale d'Angers.* Soc. Angers, 5. Angers, 1908.

Valerius Maximus. *Factorumque dictorumque memorabilium libri.* 2 vols. Ex editione Joannis Kappii. London, 1823.

Verlaguet, P.A., ed. *Cartulaire de l'abbaye de Silvanès.* Arch. Rouergue. Rodez, 1910.

Viard, Jules, ed. *Les Journaux de trésor de Philippe IV le Bel.* Collection de documents inédits. Paris, 1940.

Vidal, Auguste, ed. *Comptes consulaires d'Albi (1359–1360).* Bibliothèque Méridionale, 1st ser., 5. Toulouse and Paris, 1900.

———, ed. "Costumas del Pont de Tarn d'Albi." *Rev. des Langues Romanes* 44 (1901), 481–513.

———. *Douze Comptes consulaires d'Albi du XIVe siècle.* 2 vols. Arch. historiques de l'Albigeois, fasc. 8, 9. Albi, 1906–1911.

Vignat, G., ed. *Cartulaire de l'abbaye de Notre-Dame de Baugency.* 2 vols. Mém. Orléanais, 16, 18. Orleans, 1879–1887.

Villard de Honnecourt. *The Sketchbook.* Ed. Theodore Bowie. Bloomington and London, 1959.

III. Archival Registers and Topographical and Other Dictionaries

Agen. Bosvieux and G. Tholin. *Inventaire-sommaire des archives communales antérieures à 1790. Ville d'Agen.* Paris, 1884.

Ain. Edouard Philipon. *Dictionnaire topographique du département de l'Ain comprenant les noms de lieu anciens et modernes.* Paris, 1911.

Aisne. Auguste Matton. *Dictionnaire topographique du département de l'Aisne comprenant les noms de lieu anciens et modernes.* Paris, 1871.

Amiens. Georges Durand. *Invenatire-sommaire des archives communales antérieures à 1790. Département de la Somme. Ville d'Amiens.* 7 vols. Ser. AA–FF. Amiens, 1891–1925.

Aube. Théophile Boutiot and Emile Socard. *Dictionnaire topographique du département de l'Aube comprenant les noms de lieu anciens et modernes.* Paris, 1874.

Aude. Abbé Sabarthès. *Dictionnaire topographique du département de l'Aude comprenant les noms de lieu anciens et modernes.* Paris, 1912.

Aude. A. Sabarthès and Joseph Poux. *Inventaire-sommaire des archives départementales antérieures à 1790. Aude.* Vol. 4, *Archives ecclésiastiques.* Ser. G and H. Additions. Carcassonne, 1925.

Aveyron. M.H. Affre. *Inventaire-sommaire des archives départementales antérieures à 1790. Aveyron.* Archives civiles. Ser. B. C., D, E. Vols. 1, 2. Paris, 1866–1877.

Basses-Pyrénées. Paul Raymond. *Dictionnaire topographique du département des Basses-Pyrénées.* Paris, 1863.

Beauvais. Bernard Rose. *Inventaire-sommaire des archives communales antérieures à 1790. Département de l'Oise. Ville de Beauvais.* Beauvais, 1887.

Calvados. C. Hippeau. *Dictionnaire topographique du département du Calvados comprenant les noms de lieu anciens et modernes.* Paris, 1883.

Cantal. Emil Amé. *Dictionnaire topographique du département du Cantal comprenant les noms de lieu anciens et modernes.* Soc. Auvergne. Paris, 1897.

Castres. Estadieu. *Invenatire-sommaire des archives communales de la ville de Castres (Tarn) antérieures à 1790.* Castres, 1881.

Cher. Hippolyte Boyer and Robert Latouche. *Dictionnaire topographique du département du Cher comprenant les noms de lieu anciens et modernes.* Paris, 1926.

Clermont-Ferrand. Teilhard de Chardin. *Inventaire-sommaire des archives communales antérieures à 1790. Ville de Clermont-Ferrand. Fonds de Montferrand.* 2 vols. Clermont-Ferrand, 1902–1922.

Côte-d'Or. Alphonse Roserot. *Dictionnaire topographique du département de la Côte-d'Or comprenant les noms de lieu anciens et modernes.* Paris, 1924.

Dijon. De Gouvenain. *Inventaire-sommaire des archives communales antérieures à 1790. Ville de Dijon.* 2 vols. Dijon, 1867–1883.

Dordogne. M. le vicomte de Gourgues. *Dictionnaire topographique du département de la Dordogne comprenant les noms de lieu anciens et modernes.* Soc. Dordogne. Paris, 1873.

Drôme. J. Brun-Durand. *Dictionnaire topographique du département de la Drôme comprenant les noms de lieu anciens et modernes.* Paris, 1891.

Drôme. M.A. Lacroix. *Inventaire-sommaire des archives départementales antérieures à 1790. Drôme. Archives civiles.* Series A, B, C, D. 8 vols. Vol. 8 by M.A. Lacroix and C. Fauré. 1865–1910.

Eure. Le Marquis de Blosseville. *Dictionnaire topographique du département de l'Eure comprenant les noms de lieu anciens et modernes.* Paris, 1877.

Eure-et-Loir. Lucien Merlet. *Dictionnaire topographique d'Eure-et-Loir comprenant les noms de lieu anciens et modernes.* Soc. Eure-et-Loir. Paris, 1861.

Eure-et-Loir. M.L. Merlet and M.R. *Inventaire-sommaire des archives départementales antérieures à 1790. Archives ecclésiastiques.* Ser. G, I. Vols. 6, 7. Chartres, 1890–1894.

Gard. M.E. Germer-Durand. *Dictionnaire topographique du département du Gard comprenant les noms de lieu anciens et modernes.* Paris, 1868.

Godefroy, Frédéric. *Dictionnaire de l'ancienne langue française et de tous ses dialectes du IXe au XVe siècle.* 10 vols. Paris, 1880–1902.

Grenoble. A. Prudhomme. *Inventaire-sommaire des archives communales antérieures à 1790. Ville de Grenoble.* Ser. AA–II. 5 vols. Grenoble, 1886–1924.

Haute-Loire. Augustin Chassaing and Antoine Jacotin. *Dictionnaire topographique du département de la Haute-Loire comprenant les noms de lieu anciens et modernes.* Paris, 1907.

Haute-Loire. Antoine Jacotin and Etienne Delacambre. *Inventaire-sommaire des archives départementales antérieures à 1790. Haute-Loire.* Archives ecclésiastiques. Ser. G. Clergé séculier and supplement. 2 vols. Le Puy and Yssingeaux, 1948.

Haute-Marne. Alphonse Roserot. *Dictionnaire topographique du département de la Haute-Marne comprenant les noms de lieu anciens et modernes.* Paris, 1903.

Hérault. Eugène Thomas. *Dictionnaire topographique du département de l'Hérault comprenant les noms de lieu anciens et modernes.* Paris, 1865.

Indre-et-Loire. Charles Loizeau de Grandmaison. *Inventaire-sommaire des archives départementales antérieures à 1790.* Vol. 3 *Archives ecclésiastiques.* Ser. H. Clergé régulier. Tours, 1891.

Isère. Ulysse Chevalier. *Dictionnaire topographique du département de l'Isère comprenant les noms de lieu anciens et modernes d'après les manuscrits d'Emmanuel Pilot de Thorey.* Romans, 1921.

Joanne, Paul. *Dictionnaire géographique et administratif de France.* 7 vols. Paris, 1905.

Loire-Inférieure. H. Quilgars. *Dictionnaire topographique de la Loire-Inférieure comprenant les noms de lieu anciens et modernes.* Nantes, 1906.

Lyons. M.F. Rolle. *Inventaire-sommaire des archives communales. Ville de Lyon.* Vol. 2. Paris, 1875.

Marne. Auguste Longnon. *Dictionnaire topographique du département de la Marne comprenant les noms de lieu anciens et modernes.* Paris, 1891.

Marseilles. J.A.B. Mortreuil. *Dictionnaire topographique de l'arrondissement de Marseille (Bouches-du-Rhône) comprenant les noms anciens et modernes.* Marseilles, 1872.

Marseilles. Ph. Mabilly. *Inventaire-sommaire des archives communales antérieures à 1790. Département des Bouches-du-Rhône. Ville de Marseilles.* 2 vols. Ser. AA and BB. Marseilles, 1907–1909.

Mayenne. A Angot. *Dictionnaire historique, topographique et biographique de la Mayenne.* 4 vols. Laval, 1900–1910.

Mayenne. Léon Maitre. *Dictionnaire topographique du département de la Mayenne comprenant les noms de lieu anciens et modernes.* Paris, 1878.

Meurthe. Henri Lepage. *Dictionnaire topographique du département de la Meurthe.* Société d'archéologie lorraine. Paris, 1862.

Morbihan. M. Rosenzweig. *Dictionnaire topographique du département du Morbihan comprenant les noms de lieu anciens et modernes.* Société polymathique du Morbihan. Paris, 1870.

Moselle. M. de Bouteiller. *Dictionnaire topographique de l'ancien département de la Moselle.* Paris, 1874.

Nièvre. Georges de Soultrait. *Dictionnaire topographique du département de la Nièvre comprenant les noms de lieu anciens et modernes.* Soc. Nivernaise. Paris, 1865.

Orange. L. Duhamel. *Inventaire-sommaire des archives municipales antérieures à 1790 de la ville d'Orange.* Vol. 1. Ser. AA–CC. Orange, 1917.

Orleans. Paul Veyrier, du Mauraud, François Bonnardot, Jules Rimasson, Jules Doinel, Camille Bloch and Jacques Soyer. *Inventaire-sommaire des archives communales antérieures à 1790. Département de Loiret. Ville d'Orléans.* 2 vols. Ser. AA–II. Orleans, 1907–1920.

Oxford English Dictionary, The. Ed. James A.H. Murray. 2 vols. Oxford, 1933.

Pas-de-Calais. Le Comte de Losne. *Dictionnaire topographique du département du Pas-de-Calais comprenant les noms de lieu anciens et modernes.* Paris, 1907.

Riom. François Boyer. *Inventaire-sommaire des archives communales antérieures à 1790. Département de Puy-de-Dôme. Ville de Riom.* Riom, 1892.

Romorantin. Fernand Bournon. *Inventaire-sommaire des archives communales antérieures à 1790. Département de Loir-et-Cher. Ville de Romorantin.* Blois, 1884.

Sarthe. Robert Latouche. *Dictionnaire topographique du département de la Sarthe comprenant les noms de lieu anciens et modernes.* Paris, 1950.

Savoie. J.J. Vernier. *Dictionnaire topographique. Département de la Savoie.* Chambéry, 1896.

Seine-et-Marne. Henry Stein and Jean Hubert. *Dictionnaire topographique du departement de Seine-et-Marne comprenant les noms anciennes et modernes.* Paris, 1954.

Tarn-et-Garonne. Maisonobé, Georges Bourbon, Charles Dumas de Rauly and René Toujas. *Inventaire-sommaire des archives départementales antérieures à 1790. Tarn-et-Garonne.* Ser. A, E, G, H. 3 vols. Montauban, 1894–1951.

Toulouse. Ernest Rochach. *Inventaire des archives communales antériueres à 1790. Ville de Toulouse.* Vol. 1. Ser. AA. Toulouse, 1891.

Valence. André Lacroix. *Inventaire-sommaire des archives communales et des archives hospitalières de la ville de Valence antérieures à la revolution et inventaire-sommaire des archives communales de Die et Montélimar.* Valence, 1914.

Vienne. L. Redet. *Dictionnaire topographique de la Vienne comprenant les noms de lieu anciens et modernes.* Paris, 1881.

Vivien de Saint-Martin, M. *Nouveau Dictionnaire de géographie universelle.* 7 vols. Paris, 1879–1900.

Vosges. Paul Marichal. *Dictionnaire topographique du département des Vosges comprenant les noms de lieu anciens et modernes.* Paris, 1941.

Yonne. Maximilien Quantin. *Dictionnaire topographique du département de l'Yonne comprenant les noms de lieu anciens et modernes.* Soc. Yonne. Paris, 1862.

IV. Secondary Works

Adigard des Gautries, J. "Les Noms de lieux attestés de l'Eure entre 911 et 1066." *Annales de Normandie* 4 (1954), 39–59, 237–255; 5 (1955), 15–33.

Art de vérifier les dates, L'. Par un religieux Bénédectin de la Congrégation de S. Maur. 3 vols. 3rd ed. Paris, 1783–1787.

Astre, G. "Techniques médiévales et modernes: les matériaux du pont médiéval de la Daurade." *Annales du Midi* 63 (1951), 349–354.

Auriac, Eugène d'. *Histoire de l'ancienne cathédrale et évêques d'Alby, depuis les premiers temps connus jusqu'à la fondation de la nouvelle église Sainte-Cécile.* Rapport . . . sur quelques documents inédits relatifs à l'histoire de l'ancien évêque et de la cathédrale d'Alby. Paris, 1858.

Aymard, M. "Notice sur les travaux éxécutés au pont Saint-Esprit pour la construction d'une arche marinière." *Annales des Ponts et Chaussees*, 3rd ser. (1859), pp. 1–48.

Baluze, Etienne. *Histoire généalogique de la maison d'Auvergne, justifiée par chartres, titres, histoires anciennes, & autres preuves authentiques.* 2 vols. Paris, 1708.

_____. *Vitae paparum Avenionensium. Hoc est historia pontificum Romanorum qui in Gallia sederunt ab anno Christo MCCCV usque ad annum MCCCXCIV.* New edition by G. Mollat. Vol. 1. Paris, 1914.

Bartier, A. "La Traversée de la Saône à Chalon." *Annales de Bourgogne* 26 (1954), 129.

Becker, M. Janet. *Rochester Bridge, 1387–1856: A History of Its Early Years, compiled from the warden's accounts.* London, 1930.

Bennett, Richard, and John Elton. *History of Corn Milling.* 4 vols. London and Liverpool, 1898–1904.

Berger, Elie. "Description de Paris vers 1175 par Guy de Bazoches." *Bull. Paris* 4 (1877), 38–40.

Bergevin, Louis, and A. Dupré. *Histoire de Blois.* 2 vols. Blois, 1846–1847.

Berty, Adolphe. "Recherches sur l'origine et la situation du Grand Pont de Paris, du Pont-aux-Changeurs, du Pont-aux-Meuniers, et de celui de Charles-le-Chauve." *Rev. arch.* 12 (1855), 193–220.

Bienvenue, J.M. "Recherches sur les péages angevins aux XIe et XIIe siècles." *Le Moyen-Age* 63 (1957), 208–240, 436–467.

Bisson, Thomas N. "An Early Provincial Assembly: The General Court of the Agenais in the Thirteenth Century." *Speculum* 36 (1961), 254–281.

Black, Archibald. *The Story of Bridges.* New York and London, 1936.

Blin, Léon. "Le Grand Chemin de Paris à Lyon par la vallée de la Loire au bas moyen âge (de Décize à Marcigny par la rive gauche)." *Comité des Travaux hist.* (1958), pp. 237–265.

Boissonade, Prosper. *Essai sur l'organisation du travail en Poitou, depuis le XIe siècle jusqu'à la révolution.* 2 vols. Paris, 1900.

Bonamy, M. "Mémoire sur l'inondation de la Seine au mois de décembre 1740, comparée aux inondations précédentes, avec des remarques sur l'élévation du sol de cette ville." *Mém. Acad. Inscriptions* 17 (1741–1743), 675–708.

Bonnard, Louis. *La Navigation intérieure de la Gaule à l'époque gallo-romaine.* Paris, 1913.

Borrelli de Serres, Colonel. "L'Agrandissement du Palais de la Cité sous Philippe le Bel." *Mém. Paris* 38 (1911), 1–106.

Bouché, Honoré. *La Chorographie ou description de Provence et l'histoire chronologique du même pays.* 2 vols. Paris, 1736.

Boucher de Molandon, Remi, and Adalbert de Beaucorps, *L'Armée anglaise vaincue par Jeanne d'Arc sous les murs d'Orléans: documents inédits et plan.* Extract from *Mém. Orleanais*, vol. 23. Orleans, 1892.

Bour, R.S. "Metz. Notes sur la topographie de la partie orientale del la ville." *Annuaire Lorraine* 41 (1932), 1–180.

Bourgoüin, Etienne. "La Navigation commerciale sur la basse Loire au milieu du XIVe siècle d'après un compte de péage inédit." *Revue historique*, 60e année, vol. 175 (1935), 482–496.

Bourmont, A. de. "Les Ponts de Vaas." *Rev. Maine* 22 (1887), 172–175.

Bourquelot, Felix. *Etudes sur les foires de Champagne, sur la nature, l'étendue et les règles du commerce qui s'y faisait aux XIIe, XIIIe, et XIVe siècles.* Extract from *Mém. Acad. Inscriptions*, Ser. 2, vol. 5. Paris, 1865.

Bousquié, Georges. "Histoire du pont de Saint-Maur." *Mém. Fed. Paris* 4 (1952), 285–332.

Boutiot, Théophile. *Histoire de la ville de Troyes et de la Champagne méridionale.* 4 vols. Troyes and Paris, 1870–1880.

Bouton, André. *Les Voies Antiques; les grands chemins médiévaux et les routes royales du Haute-Marne, Département de la Sarthe.* Le Mans, 1947.

Bouvier, P. "Amendes prononcées par le prévôté et le bailliage d'Orléans (1428–1429)." *Bull. Orléanais* 16 (1911–1913), 196–200.

Boyer, Marjorie N. "The Bridgebuilding Brotherhoods." *Speculum* 39 (1964), 635–650.

_____. "Rebuilding the Bridge at Albi, 1408–1410." *Technology and Culture* 7 (1966), 495–503.

_____. "Rebuilding the Bridge at Orleans (1386–1436)." *Actes du XIe Congrès internationale d'histoire des sciences, Warsovie-Cracovie 24–31 août 1965*, 6:126–128. Wraclaw, Warsaw, and Cracow, 1968.

Brangwyn, Frank, and Walter Shaw Sparrow. *A Book of Bridges.* New York, London, 1920.

Braun, Georg, and F. Hogenberg. *Civitates orbis terrarum, 1572–1618.* Edited by R.A. Skelton. Amsterdam, 1965.

Bretaudeau, Athanase Augustin. *Histoire des Ponts-de-Cé.* Supplement to *Mém. Angers*, ser. 5, vols. 4–6. Angers, 1901–1903.

Brogan, Olwen. *Roman Gaul.* With drawings by Edgar Holloway. London, 1953.

Brøndsted, Johannes. *The Vikings.* First pub. 1960. Trans. Kalle Skov. Baltimore, 1965.

Bruguier-Roure, Louis. *Les Constructeurs de ponts au moyen-âge. Récits légendaires ou historiques suivis de la description des ponts remarquable bâtis aux XIIe et XIIIe siècles.* Extract from *Bulletin monumentale.* Paris, 1875.

Calmet, Dom Augustin. *Histoire ecclésiastique et civile de Lorraine.* 3 vols. Nancy, 1728.

Cambridge Economic History. Vol. 2. Edited by M. Postan and E.E. Rich. Cambridge, 1936.

Catel, Guillaume. *Histoire des comtes de Tolose.* Toulouse, 1623.

Cazelles, Raymond. *Nouvelle Histoire de Paris de la fin du règne de Philippe August à la mort de Charles V, 1223–1380.* Paris, 1972.

Chalande, J. "La Première Arche du Pont-Vieux, Toulouse." *Bull. Midi*, ser. 2, no. 42–43 (1912–1914), 327–330.

_____. "Les Formations alluviales dans le bassin de la Garonne à Toulouse depuis le douzième siècle." *Mém. Acad. Toulouse*, ser. 10, 12 (1912), 65–80.

Champion, Maurice. *Les Inondations en France depuis le VIe siècle jusqu'à nos jours. Recherches et documents. . . .* 6 vols. Paris, 1858–1864.

Champion, Pierre. *La Vie de Charles d'Orléans (1394–1465).* Bibliothèque du XVe siècle, 13. Paris, 1911.

Champollion-Figeac, Aimé. *Droits et usages concernant les travaux de construction, publics ou privés sous la troisième race des rois de France...(De l'an 987 à l'an 1380) d'après les chartes et autres documents originaux.* Extract from *Rev. arch.* Paris, 1860.

Chartrou, Josèphe. *L'Anjou de 1109 à 1151. Foulque de Jérusalem et Geoffroi Plantagenet.* Paris, 1928.

Châteaubriand, François August René, vicomte de. *Génie du Christianisme et défense du génie du Christianisme.* 2 vols. Paris, n.d.

Châtelain, Abel. "Les Ponts du Rhône. Etudes de géographie humaine." *Les Etudes Rhodaniennes. Revue de géographie regional* 19 (1944), 109–139.

Chenesseau, G. "Note sur un des piles de l'ancien pont d'Orléans." *Bull. Orléanais* 24 (1943), 473 ff.

Chevalier, Ulysse. "Notice historique sur le pont de Romans." *Bull. Drôme* 2 (1867), 308–332.

Clouzot, E. "Les Inondations à Paris du VIe au XXe siècle." *La Géographie* 23 (1911), 81–100.

Collin, Alexandre. "Existe-t-il des vestiges apparents d'un pont dans le lit de la Loire en face de Gien-le-Vaux?" *Mém. Orléanais* 19 (1866), 252–290.

———. *Le Pont des Tourelles à Orléans (1120–1760). Etude sur les ponts au moyen-âge. Mém. Orléanais*, 26, 27. Orleans, 1895.

Corrozet, Gilles. *Les Antiquitez, chroniques, et singularitez de Paris, ville capitale du Royaume de France. . . .* 2nd ed. Paris, 1561.

Cros-Mayrevieilh, [Jean Pierre]. *Les Monuments de Caracssonne.* Paris, 1850.

Crozet, René. "Recherches sur les ponts du Moyen Age en Haute-Poitou." *Bull. Ouest*, 4th ser., 10 (1970), 501–524.

Dartein, Fernand de. *Etudes sur les ponts en pierre remarquables par leur décoration antérieures au XIXe siècle.* Vol. 1. Paris, 1912.

Daumas, Maurice, ed. *Histoire générale des techniques.* Vol. 1, *Les Origines de la civilisation technique.* Paris, 1962.

Davey, Norman. *A History of Building Materials.* London, 1961.

Deffontaines, Pierre. *Les Hommes et leurs travaux dans le pays de la moyenne Garonne (Agenais, Bas-Quercy).* Facultés catholiques de Lille, Mém. et travaux, fasc. 39. Lille, 1932.

Degrand, E. *Construction: Notions historiques, fondations, ponts et viaducs au dessus de l'etiage.* Vol. 2 of Degrand and Jean Résal, *Ponts en maçonnerie.* Paris, 1888.

Delamare, Nicolas. *Traité de la police, où l'on trouvera l'histoire de son établissement, les fonctions et prérogatives de ses magistrats, toutes les loix et tous les reglemens qui la concernent.* 4 vols. Paris, 1705–1738. Vol. 4 by Le Cler du Brillet.

Delisle, Léopold, and Louis Passy, eds. *Mémoires et notes de August Le Prevost pour servir à l'histoire du département de l'Eure.* Soc. Eure, 2. Evreux, 1864.

De Martonne, A. "Notice historique sur l'ancien pont de Blois et sa chapelle." *Mém. Orléanais* 6 (1863), 415–443.

Denifle, Heinrich. *La Desolation des églises, monastères et hôpitaux en France pendant la querre de cent ans.* 2 vols. Paris, 1897–1899.

Desjardins, Ernest. *Géographie historique et administrative de la Gaule romaine.* 4 vols. Paris, 1876–1893.

Dessalles, L. "Rançon du roi Jean." *Mélanges de littérature et d'histoire*, pp. 145–321. Published by Soc. des bibliophiles français. Paris, 1850.

Devals aîné [Jean Ursule]. *Etudes historiques et archéologiques sur le département de Tarn-et-Garonne.* Caen, 1866.

Devic, Dom Claude, and Dom J. Vaissète. *Histoire générale de Languedoc.* New edition by Edouard Mabille, et al. 15 vols. Toulouse, 1872–1893.

Deville, A. "Recherches sur l'ancien pont de Rouen, lues à la séance du 10 décembre 1830, précis analytique." *Travaux Acad. Rouen* (1831), pp. 166–173.

Didron aîné [Adolphe Napoléon]. "Architecture civile du moyen-âge. Le Pont de Montauban." *Annales arch.* 16 (1856), 39–48.

Dion, Roger. *Le Val de Loire, étude de géographie régionale.* Tours, 1934.

Douais, C. "Des Fortunes commerciales à Toulouse et de la topographie des églises et maisons religieuses de Toulouse d'après deux testaments (XIIIe–XIVe siècles). *Mém. Midi* 15 (1891), 24–51.

Doublet, Jacques. *Histoire de l'abbaye de S. Denys . . . contenant les antiquitez d'icelle, les fondations, prerogatives, et privileges. . . .* Paris, 1625.

Du Breul, Jacques. *Le Theatre des antiquitez de Paris ou est traicté de la fondation des églises et chapelles de la cité, université, ville et diocèse de Paris.* Paris, 1612.

Dulaure, Jacques Antoine. *Histoire de Paris depuis les premiers temps historiques.* Continuée jusqu'à nos jours par Camille Leynadier. Paris, 1852.

Du Mège, A.L.C.A. *Histoire des institutions religieuses, politiques, judiciares et littéraires de la ville de Toulouse.* 4 vols. Toulouse, 1844–1846.

Duplomb, Charles. *Histoire générale des ponts de Paris.* 2 vols. Paris, 1911–1913.

Dupré, Guy. *Un Pont au moyen-âge. Le Pont de Saint-Esprit suivi d'une étude sur l'impôt du sel: "Le Petit blanc."* Pont-Saint-Esprit, 1947.

Durand, Charles. "L'Ermitage du pont de la Cité." *Bull. Périgord* 43 (1916), 229–230.

Easton, Stewart Copinger, and Helene Wieruszowski. *The Era of Charlemagne: Frankish State and Society.* Princeton, 1961.

Egbert, Virginia Wylie. *On the Bridges of Mediaeval Paris: A Record of Early Fourteenth-Century Life.* Princeton, 1974.

Emerson, William, and Georges Gromort. *Old Bridges of France.* Preface by Victor Laloux, watercolors by Pierre Vignal, drawings by Louis G. Rosenberg and Samuel Chamberlain. New York, 1925.

Emerton, Ephraim. "Altopascio–A Forgotten Order." *American Historical Review* 29 (1923), 1–23.

Emery, Richard W. *The Jews of Perpignan in the Thirteenth Century. An Economic Study Based on Notarial Records.* New York, 1959.

Enlart, Camille. *Manuel d'archéologie française depuis les temps mérovingiens jusqu'à la Renaissance.* 3 vols. in 5. Paris, 1904–1929.

Fabre, Augustin. *Histoire des hôpitaux et des institutions de bienfaisance de Marseille.* 2 vols. Marseille, 1854–1855.

Fagniez, Gustave. *Etudes sur l'industrie et la classe industrielle à Paris au XIIIe et au XIVe siècle.* Bibl. Ecole Hautes-Etudes, 33. Paris, 1877.

Fallières, O. "Le Pont d'Agen en 1381." *Congrès arch. de France tenu à Agen et Auch, en 1901*, 68th session, pp. 433–441.

Falque, Maurice. *Le Procès du Rhône et les contestations sur la propriété d'Avignon (1302–1818).* Recherches historiques et documents sur Avignon, le Comtat-Venaissin et le principauté d'Orange, 2. Paris, 1908.

Farcy, Paul de. "Aveux de Châteaugontier aux XVe et XVIIe siècles." *Bull. Mayenne* 13 (1897), 249–285.

Favre, Edouard. *Eudes, comte de Paris et roi de France (882–898).* Annales de l'histoire de France à l'époque carolingienne. Bibl. Ecole Hautes-Etudes, 99. Paris, 1893.

Félibien, Michel, and Guy-Alexis Lobineau. *Histoire de la ville de Paris.* 5 vols. Paris, 1725.

Fliche, Augustin. *Le Règne de Philippe Ier, roi de France (1060–1108).* Paris, 1912.

F.L.M. "Le Pont Saint-Bénézet." *Bull. Vaucluse* 5 (1883), 20–27, 52–58, 86–101, 499–509.

Fontenay, Harold de. *Autun et ses monuments avec un précis historique par Anatole de Charmasse.* Autun, 1889.

Fournier, Paul. *Le Royaume d'Arles et de Vienne (1138–1378). Etude sur la formation territoriale de la France dans l'Est et Sud-Est.* Paris, 1891.

Fowler, W. Warde. *The Religious Experience of the Roman People from the Earliest Times to the Age of Augustus.* London, 1922.

Fréville, Ernest de. *Mémoire sur le commerce maritime de Rouen depuis les temps les plus reculés jusqu'à la fin du XVIe siècle.* 2 vols. Rouen and Paris, 1857.

Galabert, F. "Archives des hospices de Toulouse, notice de documents principaux." *Bull. Midi*, 2nd ser., 44–45 (1919), 355–360.

Gallia Christiana. Begun by the Benedictines of St. Maur and continued by the Académie des Inscriptions. 16 vols. Paris, 1717–1765.

Gauthier, Léon. *Les Lombards dans les deux Bourgognes.* Bibl. Ecole Haute-Etudes, 156. Paris, 1908.

Gautier, Hubert. *Traité des ponts ou il est parlé de ceux des Romains et de ceux des modernes.* Vol. 1. First published 1716. 4th ed. Paris, 1765.

Gayet, Louis. *Monographie de trois monastères, fortifiées du XIVe siècle, construits au sudest d'Avignon: Montfavet, par le cardinal Bertrand de Montfavet; Ste Praxède d'Espagne, par le cardinal Pierre Gomez de Barrosso; Chartreuse de Bonpas, par le cardinal Simon de Langham.* Avignon, 1886.

Géraud, P.H.J.F. *Paris sous Philippe le Bel d'après des documents originaux, et notamment d'après un manuscrit contenant le rôle de la taille imposée sur les habitants de Paris en 1292.* Paris, 1837.

Germer-Durand, E. "Le Prieuré et le pont de Saint-Nicolas de Campagnac. Fragment d'histoire locale." *Mém. Acad. Gard* 27 (1863), 137–313.

Gies, Joseph. *Bridges and Men.* Garden City, N.Y., 1963.

Gille, Bertrand. "Le Moulin à eau: une révolution technique médiévale." *Techniques et civilisations* 3 (1954), 1–15.

Gilliard, Charles. "Problèmes d'histoire routière. I. L'Ouverture de Gothard." *Annales: Economies, Sociétés, Civilisations* 1 (1929), 177–182.

Girard, Joseph. *Evocation du vieil Avignon.* Paris, 1958.

Girardot, Bon de. "Des Ponts au XIIIe siècle." *Annales arch.* 7 (1847), 17–25.

Giraud, Paul Emile. *Essai historique sur l'abbaye de S. Bernard et sur la ville de Romans. Accompagnée de pièces justificatives inédites, entre autres du cartulaire de Romans.* 5 vols. in 3. Lyons, 1856–1869.

"Glanures historiques." *Rev. Tarn* 7 (1888–1889), 240; 8 (1890–1891), 212.

Grégoire, Henri. *Recherches historiques sur les congrégations hospitalières des frères pontifes ou constructeurs de ponts.* Paris, 1818.

Grenier, Albert. *Manuel d'archéologie gallo-romaine.* Deuxième partie, *L'Archéologie du sol. Les Routes.* Vol. 6 in J. Déchelette, *Manuel d'archéologie préhistoire celtique et gallo-romaine.* Paris, 1934.

Gröhler, Hermann. *Über Ursprung und Bedeutung der französischen Ortsnamen.* 2 vols. Heidelberg, 1913–1933.

Guérout, Jean. "Le Palais de la Cité à Paris, des origines à 1417. Essai topographique et archéologique." *Mém. Fed. Paris* 1 (1949), 57–212; 2 (1950), 21–204.

Guides Bleus, see Monmarché.

Guigue, M.–C. "Notre-Dame de Lyon: recherche sur l'origine du pont de la Guillotière et du grand Hôtel-Dieu et sur l'emplacement de l'hôpital fondé à Lyon, au VIe siècle, par le roi Childebert et la reine Ultrogothe." *Mém. Lyon* (1874–1875), pp. 227–365.

______, and Georges Guigue. "Origine du Pont-de-la-Guillotière à Lyon." In their *Bibliothèque historique du Lyonnais,* 1:128–131. Lyons, 1886.

Haitze, Pierre-Joseph de. *Histoire de S. Bénézet, entrepreneur du pont d'Avignon. Contenant celle de l'ordre des religieux pontifes, par Magne Agricol.* Aix, 1708.

Halphen, Louis. *Paris sous les premiers Capétiens (987–1223). Etude de topographie historique.* Bibliothèque d'Histoire de Paris. Paris, 1909.

Haskins, Charles Homer. *The Renaissance of the Twelfth Century.* First pub. 1927. Cited from Meridian pb edition. Cleveland and New York, 1965.

Hélyot, Pierre. *Dictionnaire des ordres religieux ou histoire des ordres monastiques, religieux et militaires.* 4 vols. First pub. 1714–1719. Paris, 1847–1859.

Histoire littéraire de la France où l'on traite de l'origine et du progrès, de la décadence et du rétablissement des sciences. Vol. 9. By the Benedictines of the congregation of St. Maur. Paris, 1750.

Hoffman, Hartmut. *Gottesfriede und Treuga Dei.* Schriften der Monumenta Germaniae Historica, 20. Stuttgart, 1964.

Holland, Louise Adams. *Janus and the Bridge.* American Academy in Rome, Papers and Monographs, 21. Rome, 1961.

Holmes, Thomas Rice Edward. *Caesar's Conquest of Gaul.* 2nd ed. London, 1931.

Home, Gordon C. *Old London Bridge.* London, 1931.

Houth, Emile. "Catalogue des actes de Robert II, comte de Meulan." *86e Congrès des Sociétés savantes, Montpellier, 1961. Comité des Travaux hist.* (1961), pp. 499–543.

_____. "Galeran II, comte de Meulan. Catalogue de ses actes. Précédé d'une étude biographique. *Comité des Travaux hist.* (1960), pp. 627–682.

Huisman, Georges. *La Jurisdiction de la municipalité parisienne de saint Louis à Charles VII.* Bibliothèque d'Histoire de Paris. Paris, 1912.

Hurtaut, Pierre Thomas Nicolas, and Magny. *Dictionnaire historique de la ville de Paris et de ses environs.* 4 vols. Paris, 1779.

Imbert, Jean. *Histoire des hôpitaux français: contribution à l'étude des rapports de l'église et de l'état dans le domaine de l'assistance publique. Les Hôpitaux en droit canonique du décret de Gratien à la sécularisation de l'administration de l'Hôtel-Dieu de Paris en 1505.* L'Eglise et l'état au moyen-âge, 8. Paris, 1947.

Jaillot, J.B.M. *Recherches critiques, historiques et topographiques sur la ville de Paris, depuis les commencemens connus jusqu'à présent.* 5 vols. New ed. Paris, 1782.

Johnson, R.P. "Note on Some Manuscripts of the Mappae Clavicula." *Speculum* 10 (1935), 72–81.

Jolibois, E. "Le Vieux Pont d'Albi." *Rev. Tarn* 1 (1877), 73; 2 (1878), 197–204.

Jones, Gwilym Peredur. "Building in Stone in Medieval Western Europe." *Cambridge Economic History*, 2:494–512. Cambridge, 1936.

Joudou, J.B. *Avignon, son histoire, ses papes, ses monuments et ses environs.* Avignon, 1842.

Jullian, Camille Louis. *Histoire de la Gaule.* 8 vols. First pub. 1908–1926. Brussels, 1964.

Julliot, G. "Les Ponts de l'Yonne à Sens, les deux moulins qu'on devait y etablir en 1367, puis en 1546, et les moulins du Roy." *Bull. Sens* 16 (1894), 97 ff.

Jusserand, J.J. *English Wayfaring Life in the Middle Ages (Fourteenth Century).* Trans. by Lucy Toulmin Smith. 2nd ed. London, 1920.

Kleinclausz, Arthur. *Histoire de Lyon.* Vol. 1. With the collaboration of Déniau, Doucet, et al. Lyons, 1908.

Labande, L.H. *Avignon au XIIIe siècle. L'Evêque Zoen Tencarari et les Avignonais.* Paris, 1908.

_____. "Etudes d'histoire et d'archéologie Romane. Saint-Symphorien de Caumont." *Mém. Acad. Vaucluse* 19 (1900), 179–199.

_____. "Guide archéologique du congrès d'Avignon." *Congrès arch. de France, 76th session, Avignon, 1909,* 1:1–311. Paris and Caen, 1910.

Lafaille, G. *Annales de la ville de Toulouse depuis la reünion de la comté de Toulouse à la couronne.* Vol. 1. Toulouse, 1687.

Lahondès, Jules de. *Les Monuments de Toulouse: histoire, archéologie, beaux-arts.* Toulouse, 1920.

Lance, Adolphe. "Rapport relatif à la construction et à la réparation des ponts antérieurement au XVIIe siècle." *Rev. Soc. Savantes,* ser. 5, 1 (1870–1871), 124 ff.

Langlois, C.V., and H. Stein. *Les Archives de l'Histoire de France.* 3 parts. Paris, 1891–1893.

Larchey, Lorédan. "Notice historique sur l'hôpital Saint-Nicolas de Metz." *Mém. Acad. Metz* 34 (1852–1853), 173–228.

Lebeuf, Jean. *Histoire de la ville et de tout le diocèse de Paris. Rectifications et additions par F. Bournon.* 6 vols. Paris, 1883–1893.

_____. *Mémoire concernant l'histoire civile et ecclésiastique d'Auxerre et de son ancien diocèse, continués jusqu'à nos jours avec addition de nouvelles preuves et annotations par Ambroise Challe et M. Quantin.* 4 vols. 1848–1855.

Leblanc, J. "Pont du Rhône entre Vienne et Sainte-Colombe." *Congrès arch. de la France, 46e session, Vienne, 1879,* pp. 89–105. Paris, 1880.

Le Bret, Henry. *Histoire de la ville de Montauban.* 2 vols. Montauban, 1668.

Lecoy de la Marche, Albert. *Le Roi René, sa vie, son administration, ses travaux artistiques et littéraires d'après les documents inédits des archives de France et d'Italie.* 2 vols. Paris, 1875.

Lefort, F. "Histoire d'un manuscrit du XIIIe siècle relatif à la construction des premiers ponts sur le Rhône." *Travaux Acad. Reims* 76 (1884–1885), 206–227.

_____. "La Légende de Saint-Bénézet." *Rev. des Questions hist.* 23 (1878), 555–570.

_____. "Le Premier Pont construit sur le Rhône à Avignon." *Travaux Acad. Reims* 71 (1881–1882), 373–399.

Legeau, Fortuné. *Recherches historique sur Vaas et Lavergnat.* Paris, 1855.

Le Roux de Lincy, A.J.V. "De la chute et de la reconstruction du pont Notre-Dame à Paris (1498–1510)." *Bibl. Ecole Chartes* 7 (1845–1846), 32–51.

Lesourd, Paul. "Histoire des ponts de Tours." *Bull. Touraine* 10 (1895–1896), 520–539.

Lewis, Archibald R. *The Development of Southern French and Catalan Society, 718–1050.* Austin, Texas, 1965.

Loisel, Antoine. *Mémoires des pays, villes, comté et comtes, evesché et evesques, pairrie, communes et personnes de renom de Beauvais et Beauvaisis.* Paris, 1617.

Lombard-Jourdan, Anne. *Paris: Genèse de la "Ville." La rive droite de la Seine des origines à 1223.* Institut de Recherche et d'Histoire des Textes. Paris, 1976.

Longnon, August. *Atlas historique de la France depuis César jusqu'à nos jours.* 3 vols. Paris, 1885–1889.

———. *Géographie de la Gaule au VIe siècle.* Paris, 1878.

———. *Les Noms de lieu de la France. Leur origine, leur signification, leurs transformations.* Paris, 1920–1929.

Lot, F. "Mélanges Carolingiens: Suite II. Le Pont de Pîtres." *Le Moyen-Age* 18 (1905), 1–33, 127–139.

———, and R. Fawtier. *Le Premier Budget de la monarchie française. Le Compte général de 1202–1203.* Bibl. Ecole Hautes-Etudes. Paris, 1932.

Luce, Siméon. "Pièces inédites relatives à Etienne Marcel et à quelque-uns de de ses principaux adhérents." *Bibl. Ecole Chartes* 21 (1859–1860), 73–92.

Luchaire, Achille. *Etudes sur les actes de Louis VII.* Histoire des institutions monarchiques de la France sous les premiers Capétiens, Mém. et doc. First pub. 1885. Brussels, 1964.

———. *Les Premiers Capétiens (987–1137).* In Ernest Lavisse, *Histoire de France*, 2. Paris, 1901.

———. *Social France at the Time of Philip Augustus.* Trans. by E.B. Krehbiel. New York, 1912.

Mantellier, Philippe. *Histoire de la communauté des marchands frequentant la rivière de Loire et fleuves descendant en icelle.* 3 vols. Orleans, 1867–1869.

———. "Mémoire sur la valeur des principales denrées et marchandisses qui se vendaient ou se consommaient en la ville d'Orléans, au cour des XIVe, XVe, XVIe, XVIIe et XVIIIe siècles." *Mem. Orléanais* 5 (1862), 103–500.

Marnotte, P. "Mémoire sur la voie romaine qui traversait Besançon, et description des antiquités découvertes lors des fouilles pratiquées pour l'établissement des fontaines de cette ville." *Acad. Besançon* (24 August 1852), pp. 120–136.

Martin, Henry. *Légende de Saint Denis. Reproduction des miniatures du manuscrit original présenté en 1317 au roi Philippe le Long.* Paris, 1908.

Mellier, Etienne. "Les Ponts anciens et modernes sur le Rhône à Valence." *Bull. Drôme* 36 (1902), 133–153, 249–269, 395–408; 37 (1903), 89–108, 183–198, 273–288, 422–437; 38 (1904), 147–181, 421–438; 39 (1905), 315–335, 376–384.

Menestrier, Claude François. *Histoire civile ou consulaire de Lyon, justifiée par chartes, titres, chroniques, manuscrits, autheurs anciens & modernes, et autres preuves.* Lyons, 1696.

Michel, R. "Les Constructions de Jean XXII à Bonpas." *Mélanges Ecole franç. de Rome* 31 (1911), 369–385.

Millin, Aubin-Louis. *Voyage dans les départemens du Midi de la France.* 4 vols. Paris, 1807–1811.

M.L.R. [Claude-Marin Saugrain]. *Nouveau voyage de France, géographique, historique et curieux.* Paris, 1771.

Mock, Elizabeth Bauer. *The Architecture of Bridges.* New York, 1949.

Monmarché, Georges and François. *France.* Les Guides bleus. Paris, 1964.

Moranvillé, H. "Rapports à Philippe VI sur l'état de ses finances." *Bibl. Ecole Chartes* 48 (1887), 380–395.

Moreau, E. and A.G. "Une Vue de vieux pont de Laval, vers 1676, et du pont de Château-Gontier." *Bull. Mayenne*, 2nd ser., 29 (1913), 235–243.

Mundy, John H. "Charity and Social Work in Toulouse, 1100–1250." *Traditio* 22 (1966), 203–287.

———. *Liberty and Political Power in Toulouse, 1050–1230.* New York, 1954.

———. Richard W. Emery, Benjamin N. Nelson, eds. *Essays in Medieval Life and Thought Presented in Honor of Austin Patterson Evans.* New York, 1955.

Needham, Joseph. *Science and Civilisation in China.* With the collaboration of Wang Ling. Vol. 4, *Physics and Physical Technology*, Part 2, *Mechanical Engineering.* Cambridge, 1965.

Nodier, Charles, J. Taylor and Alph. de Cailleux. *Voyages pittoresques et romantiques dans l'ancienne France.* 17 vols. Paris, 1820–1878.

Pansier, P. "Les Anciens Hôpitaux d'Avignon." *Annales Avignon* 15 (1929), 5–114.

———. "Complément à l'histoire de l'ordre des frères du pont d'Avignon." *Annales Avignon* 14 (1928), 89–90.

———. "Histoire de l'ordre des frères du pont d'Avignon (1181–1310)." *Annales Avignon* 7 (1920–1921), 5–74.

———. "Guilhem Vial, fustier, fournisseur du pape et de nos seigneurs les cardinaux (1351–1388)." *Mém. Acad. Vaucluse*, 2nd ser., 7 (1907), 331–363.

———. "Note sur une bulle de Calixte III accordant des indulgences à l'oeuvre du pont." *Annales Avignon* 1 (1912), 169 ff.

Paradin, Guillaume. *Mémoires de l'histoire de Lyon.* Lyons, 1573.

Parsons, William Barclay. *Engineers and Engineering in the Renaissance.* Baltimore, 1939.

Perronet, Jean Rodolphe. *Description des projets et de la construction des ponts de Neuilly, de Mantes, d'Orléans & autres du project du canal de Bourgogne.* 2 vols. Paris, 1782–1783.

Petit-Dutaillis, Charles. *Etude sur la vie et le règne de Louis VIII (1187–1226).* Bibl. Ecole Hautes-Etudes. Paris, 1894.

Pevsner, Nikolaus. *An Outline of European Architecture.* Harmondsworth, England, 1943.

Poncelet, Albert. "Relation originale du prêtre Idon sur la translation de S. Liboire à Paderborn." *Analecta Bollandiana* 22 (1903), 146–172.

Poole, Reginald Lane. "The Masters of the Schools at Paris and Chartres in John of Salisbury's Time." *Studies in Chronology and History*, ed. Austin Lane Poole, pp. 223–247. Oxford, 1934.

Porter, Arthur Kingsley. *Lombard Architecture.* 4 vols. New Haven and London, 1915–1917.

Pottier, Le Chanoine. "Le Pont de Montauban." *Bull. Tarn-et-Garonne* 1 (1869), 33–39.

Pougeois, Alexandre. *L'Antique et royale cité de Moret-sur-Loing (Seine-et-Marne).* Paris, 1875.

Poux, Joseph. *La Cité de Carcassonne, histoire et description.* Vol. 1. Toulouse, 1922.

_____. "Une Vue de Carcassonne faussement attribuée à l'an 1467." *Acad. des Inscriptions et Belles-Lettres, Comptes rendu* (1912), pp. 182–190.

Quantin, Maximilien. "Histoire de la rivière d'Yonne." *Bull. Yonne* 39 (1885), 349–498.

Quesvers, Paul. "Les Ponts de Montereau-Fault-Yonne." *Annales Gâtinais* 5 (1887), 1–8, 73–84, 233–247; 6 (1889), 164–183.

Quicherat, J. "Le Commerce et les métiers de Paris au moyen âge." *Magasin pittoresque* 14 (1846), 217–222.

Ramelli, Agostino. *Le Divers e artificiose machine.* Paris, 1588.

Rascol, P. "Albi, étude géographique." *Rev. géog. Pyrénées* 4 (1933), 73–125, 145–199.

Rashdall, Hastings. *The Universities of Europe in the Middle Ages.* New ed. by F.M. Powicke and A.B. Emden. 3 vols. Oxford, 1936.

Raynaud, Théophile. "Sanctus Ioannes Benedictus pastor et pontifex Avenione." In *Opera omnia*, 8:135–188. First pub. 1643. Lyons, 1665.

Rey, Dieudonné. *Etudes archéologiques sur le vieux Millau*, 2, *Le Pont Vieux.* Millau, 1923.

Ricci, Corrado. *L'Architettura Romanica in Italia.* Stuttgart, 1925.

Richard, Alfred. *Histoire des comtes de Poitou, 778–1204.* 2 vols. Paris, 1903.

Richard, J. "Passage de Saône aux XIIe et XIIIe siècles." *Annales de Bourgogne* 22 (1950), 245–274.

Richard, Jules-Marie. *Une Petite-nièce de Saint Louis; Mahaut, comtesse d'Artois et de Bourgogne (1302–1329). Etude sur la vie privée, les arts et l'industrie, en Artois et à Paris au commencement du XIVe siècle.* Paris, 1887.

Robins, Frederick William. *The Story of the Bridge.* Birmingham, England, 1948.

Rossignol, Elie. "Droits de navigation dans le Tarn, à Saint-Juéry et de passage au pont de Gaillac." *Congrès scientifique de France*, 28e session, Bordeaux, 1861, 4:577–583.

Rouvet, Massilon. "Le Pont d'Avignon. Texte et dessins." Extract from *Session des Beaux-Arts, 1890*, and from *Mém. Nivernais.* Nevers, n.d.

Sagnier, A. "Le Pont Saint-Bénézet." *Congrès arch. de France, 49e session, 1882*, pp. 259–282. Avignon, 1882.

_____. "Les Ponts romains sur le Rhône." *Bull. Vaucluse* 1 (1879), 1–11, 45–55, 93–101.

Saint-Victor, Jacques Benjamin Maximilien Bins, comte de. *Tableau historique et pittoresque de Paris depuis les Gaulois jusqu'à nos jours.* 3 vols. Paris, 1808–1811.

Salzman, L.F. *Building in England down to 1540: A Documentary History.* Oxford, 1952.

Samazeuilh, Jean François. *Histoire de l'Agenais, du Condomois et du Bazadais.* 2 vols. Auch, 1846–1847.

Sautel, Joseph, and L. Imbert. *Les Villes romaines de la vallée du Rhône.* Avignon, 1926.

Suaval, Henri. *Histoire et recherches des antiquités de la ville de Paris.* 3 vols. Paris, 1724.

Schneider, Jean. *La Ville de Metz aux XIIIe et XIVe siècles.* Nancy, 1950.

Sclafert, Thérèse. *Le Haut-Dauphiné au Moyen-Age.* Paris, 1926.

Séjourné, Paul. *Grandes Voûtes.* 6 vols. Bourges, 1913–1916.

Séroux d'Agincourt, Jean Baptiste Louis Georges. *Histoire de l'art par les monumens depuis sa décadence au IVe siècle jusqu'à son renouvellement au XVIe siècle.* 4 vols. Paris, 1823.

Sicard, Germain. *Aux Origines des sociétés anonymes. Les Moulins de Toulouse au moyen âge.* Ecole pratique, VIe sec., 5. Paris, 1953.

Smith, H. Shirley. *The World's Great Bridges.* Rev. ed. New York, 1964.

Steinman, David B., and Sarah R. Watson. *Bridges and Their Builders.* 2nd ed. New York, 1957.

Tanon, L. *Histoire des justices des anciennes églises et communautés monastiques de Paris suivie des registres inédits de Saint-Maur-des-Fossés, Saint Geneviève, Saint-Germain-des-Près, et du registre de Saint-Martin-des-Champs.* Paris, 1883.

Tholin, G. "Les Ponts sur la Garonne. Extrait de l'abrégé chronologique des antiquités d'Agen par Labrunie." *Rev. Agenais* 5 (1878), 439–456.

Vacquer, Théodore. "Lettre à M. l'éditeur de la *Revue archéologique* sur la découverte d'une partie du Grand Pont de Paris bâti par Charles le Chauve." *Rev. arch.* 12 (1855), 502–507.

Valbonnais, Jean Pierre Moret de Bourchenu, marquis de. *Histoire de Dauphiné et des princes qui ont porté le nom de Dauphins, particulièrement de ceux de la troisième race.* 2 vols. Geneva, 1722.

Verneilh, F. de "Architecture civile du moyen-âge." *Annales arch.* 16 (1856), 292–299.

_____. "Architecture civile du moyen âge. Construction des ponts." *Annales arch.* 20 (1860), 98–117.

Vernet, A. "L'Inondation de 1296–1297 à Paris." *Mém. Fed. Paris* 1 (1949), 47–56.

Viard, Jules. *Documents parisiens du règne de Philippe VI de Valois (1328–1350).* 2 vols. Soc. Paris. Paris, 1899–1900.

Villepelet, F. "Le Moulin du pont de la cité, en 1607." *Bull. Périgord* 48 (1916), 181–185.

Viollet-le-Duc, Eugène Emmanuel. *Dictionnaire raisonné de l'architecture française du XIe au XVIe siècle.* 10 vols. Paris, 1854–1868.

Watson, Wilbur Jay. *Bridge Architecture, containing two hundred illustrations of the notable bridges of the world, ancient and modern, with descriptive, historical and legendary text.* New York, 1927.

_____, and Sarah R. *Bridges in History and Legend.* Cleveland, 1937.

Whitney, Charles S. *Bridges: A Study in Their Art, Science and Evolution.* New York, 1929.

Wolff, Philippe. *Commerces et marchands de Toulouse vers 1350–vers 1450.* Paris, 1954.

_____. *Histoire de Toulouse.* Toulouse, 1958.

Wünsche, August. *Der Sagenkreis der geprellten Teufel.* Vienna and Leipzig, 1905.

Index